T0298692

Intelligent Control for Modern Transportation Systems

This book comprehensively discusses concepts of artificial intelligence in green transportation systems. It further covers intelligent techniques for precise modeling of complex transportation infrastructure, forecasting and predicting traffic congestion, and intelligent control techniques for maximizing performance and safety. It further provides MATLAB® programs for artificial intelligence techniques. It discusses artificial intelligence-based approaches and technologies in controlling and operating solar photovoltaic systems to generate power for electric vehicles.

- Highlights how different technological advancements have revolutionized the transportation system.
- Presents core concepts and principles of soft computing techniques in the control and management of modern transportation systems.
- Discusses important topics such as speed control, fuel control challenges, transport infrastructure modeling, and safety analysis.
- Showcases MATLAB® programs for artificial intelligence techniques.
- Discusses roles, implementation, and approaches of different intelligent techniques in the field of transportation systems.

It will serve as an ideal text for professionals, graduate students, and academicians in the fields of electrical engineering, electronics and communication engineering, civil engineering, and computer engineering.

Intelligent Control for Modern Transportation Systems

Edited by
Arunesh Kumar Singh
Bhavnesh Kumar
Ibraheem
Asheesh Kumar Singh
Shahida Khatoon

CRC Press is an imprint of the
Taylor & Francis Group, an **informa** business

First edition published 2024
by CRC Press
2385 NW Executive Center Drive, Suite 320, Boca Raton, FL 33431

and by CRC Press
4 Park Square, Milton Park, Abingdon, Oxon, OX14 4RN

CRC Press is an imprint of Taylor & Francis Group, LLC

ISBN: 978-1-032-39300-1 (hbk)
ISBN: 978-1-032-56562-0 (pbk)
ISBN: 978-1-003-43608-9 (ebk)

DOI: 10.1201/9781003436089

Typeset in Sabon
by codeMantra

Contents

Preface

Modern transportation system is considered to be an umbrella term that covers route optimization, accident prevention, street-lightning, parking, and infrastructure applications. Understanding the importance of sustainable transportation, the United Nations Department of Economic and Social Affairs (UNDESA) has kept it under one of the 17 Sustainable Development Goals of the 2030 Agenda for Sustainable Development. With the rise of the Internet of Things (IoT) and machine learning techniques, traffic management, traveler information system, public transportation, and emergency management system have been developed. The intelligent control of modern transport systems integrates a wide range of communication, control, vehicle sensing, and electronic technologies to manage transportation problems. This book discusses the state-of-the-art technologies in the field of modern transportation systems. Special emphasis has been given to uncovering the challenges related to storage and energy management in modern transportation systems.

Today's modern world is under the significant influence of innovative technologies such as Artificial Intelligence, Deep Learning, Machine Learning, and IoT. This book aims to present the various approaches, techniques, and applications that are available in the field of modern transportation systems. It is a valuable source of knowledge for researchers, engineers, practitioners, and graduate and doctoral students working in the same field. It will also be helpful for faculty members of graduate schools and universities.

We would also like to thank the authors of the published chapters for adhering to the schedule and incorporating the review comments. We wish to extend my heartfelt acknowledgment to the authors, peer reviewers, committee members, and production staff whose diligent work shaped this volume. We especially want to thank our dedicated team of peer reviewers who volunteered for the arduous and tedious step of quality checking and critique on the submitted chapters.

Arunesh Kumar Singh, Bhavnesh Kumar, Ibraheem, Asheesh Kumar Singh, Shahida Khatoon

About the Book

Keeping in mind the importance of intelligent control techniques for the transportation system and the application of AI in this field for improvement, this book is intended to provide a collection of solutions based on artificial intelligence techniques to the readers on the challenges in modern transportation systems. This book comprehensively addresses all the aspects of AI in green transportation systems, and also each chapter follows a similar structure, so that students, teachers, and industry experts can orientate themselves within the text.

Modern transportation system is considered to be an umbrella term that covers route optimization, accident prevention, parking, and infrastructure applications. With the rise of the IoT and machine learning techniques, traffic management, traveler information system, public transportation, and emergency management system are developed. The intelligent control of modern transport systems integrates a wide range of communication, control, vehicle sensing, and electronic technologies to manage transportation problems.

Especially focusing on *Electric Vehicles* as sustainable transportation from the future perspective, this book discusses about the key barriers to their adoption like limited range, safety, and performance. Furthermore, as a prominent solution to many of these challenges, soft computing techniques have been discussed in this book to develop efficient infrastructure systems, energy management, and security systems for more user-friendly and accessible transportation systems.

About the Editors

Dr. Arunesh Kumar Singh graduated in Electrical Engineering and received an MTech in Engineering Systems from the Faculty of Engineering, Dayalbagh Educational Institute, Dayalbagh, Agra in 2002 and 2004, respectively. He received a PhD in Electrical Engineering from Jamia Millia Islamia, New Delhi in 2011. He was awarded a Postdoctoral Fellowship (PDF) for 17 months from the University of Saskatchewan under Discovery Grant-NSERC, Canada. He worked as a Jr. Telecom Officer (Executive) in BSNL from 2009 to 2011 and then joined the Department of Electrical Engineering, Jamia Millia Islamia as an Assistant Professor in February 2011. He has more than 19 years of experience in teaching and industry, and he is actively involved in guiding research scholars at UG, PG, and PhD levels. He has completed a research project sponsored by UGC and also got a project from SERB. He is a Life Member of the System Society of India (SSI), a Senior Member of IEEE, a Member of IE(I), a Professional Engineer of IE(I), and Professional Member of ACDOS (NMO IFAC). He is the recipient of BHARAT VIKAS AWARD-2018 for outstanding performance in the field of Electrical Machines & Neuro-control by ISR, Bhubaneswar. His research areas are intelligent systems, soft & quantum computing, neuro-control systems, Eddy current control systems, and neuro- and machine controls. His one patent is published in October 2021 on multi-disc ECBS. He has published various research papers in refereed journals and conferences of national and international repute.

Dr. Bhavnesh Kumar is an Assistant Professor at the Department of Instrumentation & Control Engineering, Netaji Subhas University of Technology (NSUT), Delhi. He holds a Bachelor of Technology degree in Electrical & Electronics Engineering from Uttar Pradesh Technical University, Lucknow, and a Master of Technology degree in Control & Instrumentation from Motilal Nehru National Institute of Technology, Allahabad. He has earned a PhD from Gautam Buddha University, Greater Noida in the area of artificial intelligent controllers for induction motor drives. With a teaching experience of more than 10 years, he served Gautam

Buddha University, Greater Noida, the Airports Authority of India, and KNGD Modi Engineering College, Modinagar before joining NSUT. He has published various research papers in journals and conferences. He has presented his research work at many international conferences. He is a member of various professional bodies such as the Institute of Electrical and Electronics Engineers (IEEE) and the International Association of Computer Science and Information Technology (IACSIT). His research interest is in the application of artificial intelligence to renewable energy systems and electric drives.

Prof. Ibraheem is presently working as a Professor in the Department of Electrical Engineering and Dean of the Faculty of Engineering & Technology, Jamia Millia Islamia (A Central University), New Delhi, India since 2002. He joined the Department as a Lecturer in January 1988. Before Joining Jamia Millia Islamia, he had served Delhi Development Authority for a considerable time. Dr. Ibraheem received BSc Engineering (Hons.), MSc Engineering, and PhD degrees in Electrical Engineering from Aligarh Muslim University, Aligarh, India, in 1982, 1987, and 2000, respectively. He was the Head of the Department of Electrical Engineering from 2002 to 2005. His current activities include teaching and research in the areas of power system control, optimal control theory, suboptimal control of power systems, applications of soft computing techniques in power systems, and HVDC transmission systems. Dr. Ibraheem is a member of various academic societies of national and international repute. He has been engaged continuously in guiding research activities at the graduate, postgraduate, and doctoral levels. Seven PhD degrees are already in his name as supervisor/co-supervisor, and 12 PhD research scholars are doing their research work under his guidance. He has published 115 research articles in international/national journals. He was awarded a Gold Medal from the Union Ministry of Power and Energy (India) in 1998 for one of his research articles. He chaired many sessions at national and international seminars/conferences/workshops. He got grants from AICTE, UGC, and DST for research projects. He has been an expert committee member for various committees of UGC, National Board of Accreditation, UPSC, HRD, Staff Selection Commission, etc. The Central Government appointed him as a Member of the Dargah Committee Ajmer for 5 years from 24-08-2007, and he was Vice President of the Committee.

Prof. Asheesh K Singh is a Professor in the Electrical Engineering Department at Motilal Nehru National Institute of Technology (MNNIT) Allahabad, Prayagraj (Uttar Pradesh). He received his BTech (Electrical Engineering) from HBTI Kanpur in 1991, MTech (Control Systems) from REC Kurukshetra in 1994, and his doctorate from IIT Roorkee in 2007. Since 1995, he has been on the academic staff of MNNIT Allahabad, India.

His research interests include power quality, e-mobility, smart metering, AI applications in power systems, energy management, and smart grid. He has published more than 100 research publications in peer-reviewed international journals and conferences. He has authored 1 book and 12 book chapters, and organized several short-term faculty development programs focusing on various power system issues. He has delivered over 50 lectures at various national and international forums. He has supervised 13 Doctoral and 54 MTech theses. Also, he has procured research funding of over INR 1.5 Crores since 2015 from various international and national funding agencies. He is a Senior Member of IEEE, a Life Member of the Indian Society for Technical Educational (ISTE), a Life Member of the System Society of India (SSI), and a Fellow of the Institution of Engineers (India). His active participation in IEEE activities highlights his professional services. As an IEEE member of the Uttar Pradesh Section, he volunteered as the Section Chair of the IEEE Uttar Pradesh Section from 2019 to 2020, and the Vice-chair (Sections and sub-sections) of the IEEE India Council, in the year 2021. Prof. Singh is a recipient of the Outstanding Engineer Award (OEA) – 2020 of IEEE PES UP Section Chapter, for his outstanding technical, professional, educational, and social contributions to the power engineering profession.

Prof. Shahida Khatoon obtained her BTech in Electrical Engineering from Jamia Millia Islamia in 1990 and MTech in Controls and Instrumentation from IIT Delhi in 1995. She obtained her PhD degree from Jamia Millia Islamia in 2004. Prof. Khatoon has published about 100 research papers in the area of Controls and Power System engineering in peer-reviewed international journals and conferences. Her research areas include control systems engineering, robotics and automation, soft computing techniques and their applications in power systems, control systems, and electronics engineering. Prof. Khatoon has delivered many invited lectures at various institutes and conferences. Prof. Shahida is a member of various academic societies of national and international repute. She has been Track Chair of the IEEE International Conference INDICON-2015 and IEEE INDIACOM 2020 and a technical committee member of many IEEE conferences. She has been invited as a subject expert in many public service commissions and industries. Apart from academics, she has contributed to the corporate life of the university and worked as deputy Provost from 2003 to 2008, and currently, she is serving the university as Deputy Dean, Student Welfare. Moreover, she is a reviewer for the various reputed international journals. In addition, she has also worked on a research project in AT & T Bell Lab, New Jersey, USA in 2001 and has attended many continuing education programs in North Carolina, USA during her visits in 2017, 2018, and 2019.

Contributors

Sanjeev Anand
School of Energy Management
Shri Mata Vaishno Devi University
Jammu and Kashmir

D K Chaturvedi
Faculty of Engineering
Dayalbagh Educational Institute
Agra, India

Urvashi Chauhan
Parul University
India

Himanshu Chhabra
Parul University
India

Anmol Gupta
KIET Group of Institutions
Ghaziabad, India

Rudraksh S. Gupta
Shri Mata Vaishno Devi University
Jammu and Kashmir

Prince Jain
Parul University
India

Bakul Kandpal
Department of Energy Science and Engineering
IIT Delhi, India

Amit Kukker
SIT, Symbiosis International
Pune, India

Abhishek Kumar
SOET
CMR University
Bengaluru, India

Bhavnesh Kumar
Instrumentation & Control Engineering Department
Netaji Subhas University of Technology
Delhi, India

Rohit Kumar
Electrical Engineering Department, Jamia Millia Islamia
New Delhi, India

Shubham Kumar
KIET Group of Institutions
Ghaziabad, India

Tarun Kumar
KIET Group of Institutions
Ghaziabad, India

Trisiladevi C J Nagavi
SS Science and Technology University
Mysore, India

Bhanu Pratap
National Institute of Technology
Kurukshetra, India

Shilpa Ranjan
Electrical Engineering Department
Delhi Technological University
Delhi, India

Prince Kumar Saini
National Institute of Technology
Kurukshetra, India

Shivam Sharma
KIET Group of Institutions
Ghaziabad, India

Niraj Kumar Shukla
Shambhunath Institute of Engineering and Technology
Prayagraj, India

Arunesh Kumar Singh
Electrical Engineering Department
Jamia Millia Islamia
New Delhi, India

Brijesh Singh
KIET Group of Institutions
Ghaziabad, India

Dinesh Kumar Singh
Shambhunath Institute of Engineering and Technology
Prayagraj, India

Madhusudan Singh
Electrical Engineering Department
Delhi Technological University
Delhi, India

Pavan Kumar Singh
Shambhunath Institute of Engineering and Technology
Prayagraj, India

Mini Sreejeth
Electrical Engineering Department
Delhi Technological University
Delhi, India

Arjun Tyagi
Department of Electrical Engineering
Netaji Subhas University of Technology
Delhi, India

Ashu Verma
Department of Energy Science and Engineering
IIT Delhi, India

Monika Verma
Electrical Engineering Department
Delhi Technological University
Delhi, India

M K Vismaya
JSS Science and Technology University
Mysore, India

Chapter 1

Fundamentals of modern transportation systems

Arunesh Kumar Singh and Rohit Kumar
Jamia Millia Islamia

Bhavnesh Kumar
Netaji Subhas University of Technology

D. K. Chaturvedi
Dayalbagh Educational Institute

CONTENTS

DOI: 10.1201/9781003436089-1

1.1 INTRODUCTION

The desire to learn more about one's environment is ingrained in the human race, which is why so many bright minds have sought out a method to conquer the sky. That's why the evolution of technology and associated systems is a continuous and gradual process aimed at maximising efficiency and comforts. Moving people, goods, and other items from one location to another is what we call transportation, and it can occur in many different modes of travel.

The economy of any country is directly dependent on the transportation of that country, which is the progress in transportation estimated by the size and quality of road network, rail network, and connectivity through airways and waterways. Other parameters to measure the progress in transportation are the speed of transportation, efficiency, safety, comforts, and environmental friendliness. The speed of transportation adversely affects the vehicle and personnel safety. The present mode of transportation is heavily dependent on the conventional fuel. Most policymakers are cautious about safety, pollution, and oil dependence, which are the most pressing problems caused by this increase in transportation. These problems are especially acute in the fastest-growing economies of the developing world. Reducing GHG emissions can be prioritised alongside these other transportation issues by emphasising synergies and co-benefits.

Growing economies necessitate more transportation, and better transportation links boost by facilitating trade and allowing professionals to specialise. There is still a huge gap between the number of people who have access to personal vehicles and the number of people who have access to any form of motorised transportation. Though, this is a situation that is rapidly evolving. There will be a dramatic increase in transportation-related activities over the coming decades. The demand for transportation energy is expected to rise by around 2% year, with developing nations experiencing the strongest development. By 2030, it is projected that global transportation will consume nearly 80% more energy than it does now, while also producing nearly 80% more carbon dioxide [21], if we continue to use conventional fuel. This is a major reason to shift from conventional fuel to renewable energy sources like solar energy, wind energy generating sources for battery charging for battery operated vehicles (BEVs), biofuels for conventional automotive, etc.

1.2 KEY COMPONENTS

The key components of a transportation system are the infrastructure, vehicles, and equipment, fuel, and control and communication systems [6].

1.2.1 Infrastructure

A well-developed infrastructure is necessary for successful and efficient transportation. The terminals and pathways are two examples of infrastructure. Highways, also known as roadways, are defined as paths used primarily for transporting vehicles with rubber tyres. Railways, also known as rails, are a specialised type of guided vehicle transport with predetermined track spacing (wheels). Air passages, waterways, and underground pipelines are examples of important guide ways.

1.2.2 Vehicles

Vehicles are a part of most transportation networks but not all. An example of a system that lacks this is a pipeline. When compared to the fixed components of the system (the guide way, the stations, and the terminals), the vehicles are the dynamic components. Automobiles, trucks, train locomotives, aeroplanes, and other types of vehicles abound. The materials used to build cars have an effect on the vehicles' overall size, weight, durability, and safety in the event of a collision. The trade-off between efficiency (lighter vehicles use less energy to move) and safety is a common one (lighter vehicles come out second best in a crash).

Imagine different kinds of transportation, such as cars, trains, and aeroplanes. A vehicle's motor is as individual as the vehicle itself. There are many types of vehicles that can't move on their own. Freight cars, for example, on the railroad industry, are not self-propelled and must be pulled by a locomotive. One needs a tractor unit in the trucking industry to move trailers and containers around. Specifically, we use the terms "vehicles with propulsion" and "vehicles without propulsion" to categorise various automobiles.

The power train includes the electric motors that propel the vehicles. In-line gasoline or diesel engines are examples of common power train systems. Power train systems typically cap the top speed and acceleration that can be achieved. Some power trains employ regenerative braking to recycle the energy normally lost when the vehicle slows down (decelerates) to reduce speed or stop.

1.2.3 Fuel

Fuel is another major constituent of the transportation system. Presently, coal, ethanol, methanol, diesel, natural gas, and gasoline are commonly used fuels in the transportation system. The cost and efficiency of automotive fuels are hot topics right now. The environmental effects of transportation systems are further complicated by the variety of fuels used. Many different types of renewable energy can be used to power transportation, including liquid fuel cells derived from unconventional oil, natural gas, coal, or biomass. The use of gaseous fuels such as natural gas, hydrogen, or electricity is also feasible. Several different types of feedstocks can be used to generate either hydrogen or electricity. These alternatives are all costly, and some, like liquids made from fossil fuels, can significantly increase GHG emissions without carbon sequestration.

Electrical energy is a source of power. Electricity is centrally generated, for example, from coal, nuclear or hydro power stations, normally located at remote places due to availability of resources and transmitted via high-voltage lines to enable the propulsion of automobiles by on-board electric motors. Electric vehicles (EVs), on the other hand, have batteries that are charged with electricity and then used to drive the vehicle.

Solar energy can be used to power EVs equipped with photovoltaic cells. An intriguing possibility is to use solar panels to recharge the battery of a battery-powered car while the car is in motion.

The internal combustion engine (ICE) is the lifeblood of the worldwide automotive industry. Most automobiles use gasoline-powered ICEs. Both the scarcity of oil and the growing awareness of the environmental risks it poses have spurred significant research into alternative power sources for automobiles. The development of hybrid vehicles, which use both ICE and electric motors powered by batteries, is currently underway. We may be able to cut back on our reliance on petroleum for transportation fuels thanks to advances in biofuels technology. Recent IEA research suggests that biofuels may make up 10% of transportation fuel by the year 2030. The market for ethanol as a motor fuel has been revitalised by the development of flex-fuel vehicles, which can run on any blend of gasoline and ethanol.

1.2.4 Control and communication systems

Control, communications, and positioning systems are further essential elements of transportation systems. Infrastructure, transportation networks, and vehicles all require various approaches to management. This regulating factor is typically a human being, like the vehicle's, train's, or plane's pilot. This human control element could be an air traffic controller, for example, who is not in any of the vehicles but still operates the system by giving orders to the pilots.

Technology is used as part of the control, communication, and locating systems alongside human operators. For example, one could argue that the control of the transportation system includes the use of traffic signals with their distinctive red, green, and yellow lights. The functions of both stationary and moving road signs are equivalent.

Both satellite communication and the use of global positioning system (GPS) as a location system for transportation companies are relatively new technologies that are gaining increasing importance in the transportation sector. It is possible to track down specific vehicles and provide navigational assistance to their drivers using GPS technology. For instance, vehicles could be equipped with a GPS sensor, allowing a centralised system to monitor their location and make (hopefully) impartial decisions about the most efficient route to get drivers from location A to B. Vehicle-to-grid communication is a critical component of today's transportation infrastructure.

1.3 HISTORICAL BACKGROUND

According to legend, Ferdinand Verbiest created the first steam-powered vehicle in the late 16th century [1]. The Verbiest's 2-foot-long vehicle was not designed to carry people, but rather to clear the way for future enthusiasts to build upon. While the ICE was still in its infancy, significant progress was made on the steam carriage and car in the 18th century. Karl Benz is generally regarded as the inventor of the first gasoline-powered automobile in 1885. This caused a paradigm shift in the auto industry, and in the 19th century, manufacturers shifted their focus from steam cars to those powered by gasoline engines [5]. Currently, only a select few manufacturers, such as Stanley and Board of France, produce steam cars. However, as interest in ICEs grows, steam cars like these will soon be in the minority. By 1950, only Paxton was still researching steam cars, as the trend had faded. The DobleUltimax engine, which was tested on a Ford Coupe and was said to be capable of producing 120 bhp, was never released to the public and scrapped as a result (Table 1.1).

The following key points make up a succinct history of the ICE (Table 1.2).

1.4 INFRASTRUCTURE FOR ROAD TRANSPORT: EMERGING TECHNOLOGIES

This is fixed component of a transit system, consisting of the track and related accessories, as well as stations and other operational building blocks, this system is built to last. The guide way, the framework that keeps

Table 1.1 Initial work in propelled mechanical vehicle

Year	*By*	*Notable work*
1769	Nicolas-Joseph	First working self-propelled land-based mechanical vehicle
1791	William Murdoch	Steam carriage
1815	Josef Božek	First substantiated steam car [2]

Table 1.2 History of automotive

Year	*By*	*Notable work*
1680	Christian Huygens	Internal combustion device used gunpowder as fuel.
1807	Francois Isaac de Rivaz	The first internal combustion-powered automobile was built for an internal combustion engine that ran on a fuel mixture of hydrogen and oxygen. But this concept has a serious flaw.
1858	Jean Joseph Étienne Lenoir	A dual-acting, coal-gas-fuelled internal combustion engine with electric spark ignition.They had a 50 mile road trip.
1862	Alphonse Beau de Rochas	Patented a four-stroke engine but never really built one.
1864	Siegfried Marcus	Created a 500-foot drive on rough terrain using a one-cylinder engine and a primitive carburettor coupled to a cart. Marcus created a vehicle that could run at 10 mph for a brief period of time which some historians believe was the predecessor of the modern automobile because it was the first gasoline-powered vehicle ever built.
1873	George Brayton	Unsuccessful two-stroke kerosene engine. But it was regarded as the first oil engine that was both secure and useful.
1866	EugenLangen, and Nicolaus August Otto	Created a more effective gas engine and improved on Lenoir's and de Rochas' ideas.
1876	Sir Dougald Clerk	First successful two-stroke engine
1883	EdouardDelamare	Developed a stove gas-powered single-cylinder four-stroke engine. Delamare-designs Debouteville's were quite sophisticated for the period – at least on paper, they were superior to both Daimler and Benz in certain areas. It is unclear, though, whether he really built an automobile.
1885	Gottlieb Daimler	Prototype of a modern internal combustion engine with a vertical cylinder and carburettor-injected fuel.With this engine, Daimler first created a two-wheeled vehicle called the "Reitwagen" (Riding Carriage), then a year later, the first four-wheeled automobile.
1886	Karl Benz	First gas-powered vehicle patent (DRP No. 37435).
1889	Daimler	Upgraded four-stroke engine with two V-slant cylinders and mushroom-shaped valves.
1890	Wilhelm Maybach	First 4-cylinder four-stroke engine.

vehicles stable as they move, is part of the infrastructure that varies the most between different transit systems. Despite their primary function of sustaining vertical vehicle motion, some guide ways also help regulate horizontal vehicle motion, most commonly in lateral directions across the track but occasionally in longitudinal directions as well.

1.4.1 Early road networks

As the Roman Empire expanded, it became increasingly important for armies to be able to move quickly from one region to another. However, the roads were typically muddy at the time, which greatly slowed down the passage of numerous troops. The long-lasting roads were constructed by Romans as a permanent solution to this problem. To prevent the roads from becoming muddy in clay soils, the Romans used deep roadbeds of crushed stone as an underlying layer.

1.4.2 Modern road networks

In exchange for professional upkeep, the British Parliament began enacting a series of laws beginning in the early 18th century that authorised local judges to set up toll gates along the roads. Wade's Mill which is located in England had the first working toll gate. The first non-judicial trust for the Foothill–Stony Stafford section of the London–Chester Turnpike was established by an act of Parliament in 1707.

Asphalt and concrete are the materials of choice for modern roadways. It all comes down to whether Portland cement or asphalt cement is more common in your area, but both are based on Mc Adam's theory of stone aggregate in a binder. Asphalt roads are known for their ability to "flow" slowly under the weight of traffic. Because concrete is a hard paving medium, it can withstand heavier weights, but it also costs more and needs a more carefully constructed sub-base. Consequently, asphalt is typically used for smaller roads and concrete for larger ones. Concrete roads often have a thin layer of asphalt laid on top to create a more durable surface.

1.4.3 Toll roads

A toll road is a public or private road that levies a fee (or toll) for use. It is often referred to as a turnpike or tollway. In modern times, toll roads are nearly invariably controlled-access highways. It is a sort of road pricing that is frequently applied to help pay for the costs associated with constructing and maintaining roads.

More cars on the road means more congestion in recent years. Since millions of people in developing countries take their cars to work every day instead of taking public transportation, congestion is a growing problem in these areas. An increase in the number of vehicles on the road leads to increased traffic, air pollution, and fuel wastage. Toll roads are a special category of roads that require payment before driving on them. Toll taxes are fees charged to drivers who use toll roads. The toll is only paid by those who drive on the toll road. Occasionally, more than one toll plaza may be present on a given toll road.

1.4.3.1 Electronic toll collection

A system called electronic toll collection (ETC) enables toll payments to be collected electronically, allowing for almost continuous toll collection and traffic monitoring. A sticker with an RFID (Radio Frequency Identification) chip is applied to the vehicles as part of the system, enabling automatic toll payment deduction. ETC systems are now in use around the world. Several countries around the world, including Canada, Poland, Japan, Italy, and Singapore, have adopted the ETC system. National Highways Authority of India-owned Electronic Toll Collection Systems in India that are managed by a variety of toll management companies include the NH-6 Toll Road in Kharagpur (managed by Toll Tax Toll Collection System), the Delhi Gurgaon Expressway (managed by Metro Electronic Toll Collection Systems), the Lucknow Sitapur Expressways Ltd in Uttar Pradesh (managed by Rajdeep – Toll Management System).

1.4.3.2 FASTag

Vehicle owners and drivers have traditionally stopped at the toll plaza, handed cash or credit card to the toll collector at the toll booth, and then waited for the gate to be mechanically or electronically opened before continuing on their way. Such sudden starts and stops, even on supposedly good roads, are a waste of gas. Another issue is that people have to deal with cash and wait for change.

In India, the FASTag ETC system is overseen by the National Highways Authority of India. The system was first put into place as a pilot project between Ahmedabad and Mumbai (the Golden Quadrilateral) in 2014. The Delhi–Mumbai segment of the Quadrilateral went live with the system on November 4th, 2014. In July 2015, toll booths along the Chennai-Bengaluru stretch of the Golden Quadrilateral started accepting payments via FASTag. Approximately 70% of India's toll booths, or 247, were FASTag-enabled as of April 2015. As of November 23, 2016, 95% toll booths on US highways accept FASTag payments.

A FASTag, which can be attached to a car's windshield, is required for driving through toll gates. A FASTag can be recharged in person with cash or by mail using a check, or online using a major credit card, debit card, NEFT, RTGS, or net banking. FASTag accounts can be loaded with as much as Rs 1 lakh for as little as Rs 100. Users of FASTags are required to proceed through toll booths in accordance with established protocols. There may be a designated FASTag lane at toll plazas, or the option to validate FASTags with a handheld reader. FASTag logos will be displayed on boards or electronic displays about 70 m before the toll plaza to direct drivers to the designated lanes.

1.5 VEHICLES FOR ROAD TRANSPORT: CURRENT AND FUTURE

Conventionally, heat engines power motorised vehicles on the road. Battery-powered EVs are the cutting edge of automotive technology. This means there are three main classes of automobiles. First fossil fuel-powered automobiles, second electric-powered vehicles run on batteries, and third, hybrids that combine the two power sources.

1.5.1 Conventional vehicles driven by heat engines

With the use of a heat engine, thermal energy that has been stored in a reservoir can be transformed into usable mechanical motion. Some of the heat is lost to the sink (cold body). Heat waste is an inevitable by-product of this system. ICEs and external combustion engines are the two primary categories of heat engines [3].

An ICE's combustion chamber burns a fuel (usually a fossil fuel) and an oxidiser (typically air). In an ICE, mechanical work is done when the expanding high-temperature, high-pressure gases created during combustion act on engine components like pistons, turbine blades, or nozzles.

The combustion of an external source heats an internal working fluid via the engine wall or a heat exchanger in an EC engine, a form of heat engine. The fluid creates motion and useful work by expanding and affecting the engine's mechanism. They are rarely used in today's transportation systems.

1.5.2 Internal combustion engine

In an ICE, the chemical energy in fuel is transformed into mechanical energy, which is normally made available on a spinning shaft [3]. Initially, the engine converts the fuel's chemical energy into thermal energy through combustion or oxidation with air. The high-pressure gas in an ICE is heated by the engine's combustion process, causing the mechanical parts of the engine to expand. Mechanical linkages within the engine convert this growth into the crankshaft's rotation, which is the engine's output. In order to make it easier for mechanical energy to be transferred from the source to the load, crankshafts are connected to transmissions and/or power trains. Many times, this will be the engine's source of propulsion for the vehicle.

Surface heat flow in a reciprocating engine's combustion chamber can go from zero to as high as 10 MW/m^2 and back to zero in less than 10 ms [7]. A surface's flux can vary by as much as 5 MW/m^2 from one spot to another, even though those spots may be only 1 cm apart. The complexity is compounded by the fact that the flux pattern varies greatly from cycle to cycle.

1.5.2.1 Technological advancements in ICE

Although engine emissions have decreased by almost 90% since the 1940s, they still pose a serious threat to the environment. Emissions from ICEs primarily consist of solid particles, carbon monoxide, hydrocarbons (He), and nitrogen oxides (NO_x). Hydrocarbons are fragments of fuel molecules that did not burn completely. When the air–fuel mixture is not thoroughly mixed or there is not enough oxygen to convert all the carbon to CO, carbon monoxide is produced. The exhaust from compression–ignition engines has visible solid particles, which appear as black smoke. It's not just carbon monoxide and nitrogen oxides that can be found in vehicle tailpipes; engine exhaust can also contain aldehydes, sulphur, and lead.

Technology for shrinking the size of engines has improved over time, boosting their efficiency [11]. As a result of downsizing, pumping effort is decreased since engine operating regimes are modified to accommodate increased load factors. It reduces losses due to friction as well. Researchers have begun ICE research because of the potential benefits it could bring to the transportation industry in the future, brought on by technological advancements in engine technology. The ICE of the future will likely be a multi-fuel combustion engine with a complex power train management system and adaptively controlled components.

To increase the system's efficiency, numerous attempts were made across a variety of ICE technology subsystems. An attempt to downsize the ICE and employ friction reduction techniques helps raising the efficiency of the ICE technology. There were many other production-intent initiatives that significantly increased the ICE's efficiency, such as the VCR (variable compression ratio) system, which has the ability to increase efficiency. In recent years, LTC (low-temperature combustion) has drawn interest from players in the auto industry as a viable technology to develop (Table 1.3).

1.5.3 Battery-powered vehicles

Most of the top automakers in the world have already begun mass-producing EVs due to the widespread consensus that these vehicles offer significant advantages over those powered by ICEs. According to the United Nations' sustainable development goals, EVs will become ubiquitous because of

Table 1.3 Energy efficiency capabilities [9]

ICE technology	*Efficiency increase in %*
Capacity reduction with constant power	<20
Limiting friction	<5
Cam changeover	4–10
SI engine with direct injection	2–10
Improved fuel injection technique	3–5

the positive impact they will have on the economy, the environment, and society. Being one of the cutting-edge forms of environmentally friendly transportation, EV innovation has captured the attention of many academics. On top of that, EVs are essential for reducing non-renewable resource consumption and maximising the use of renewable energy. In the Indian market, newer models such as the Mercedes-Benz EQS, Hyundai IONIQ 5, Tata Nexon, and Tata Tiago have recently been introduced.

Plug-in charging infrastructure is a standard requirement for BEVs that eliminate ICE. Electric motors powered by energy stored in a battery propel these vehicles. The propulsion systems of BEVs are entirely electric and are powered by the vehicles' batteries. Plugging the vehicle into the grid can recharge its massive battery pack, which stores the electricity used to power the vehicle. When the battery pack is fully charged, it sends power to one or more electric motors, propelling the EV. The electric motor, inverter, battery, drive train, and control module are the backbones of a fully electric vehicle. The functional block diagram of EVs is given in Figure 1.1.

1.5.3.1 Electric motor

There are many different kinds of electric motors, and they all have their own special qualities. Therefore, it is crucial to conduct an in-depth analysis of the most suitable motor for a given vehicle. Cost, low maintenance, high specific power, a simple design, and good control are just some of the features the electric motor should have in order to be used in EVs [10]. EV makers typically employ DC (direct current), induction, permanent magnet synchronous, switched reluctance, and brushless DC motor types [12].

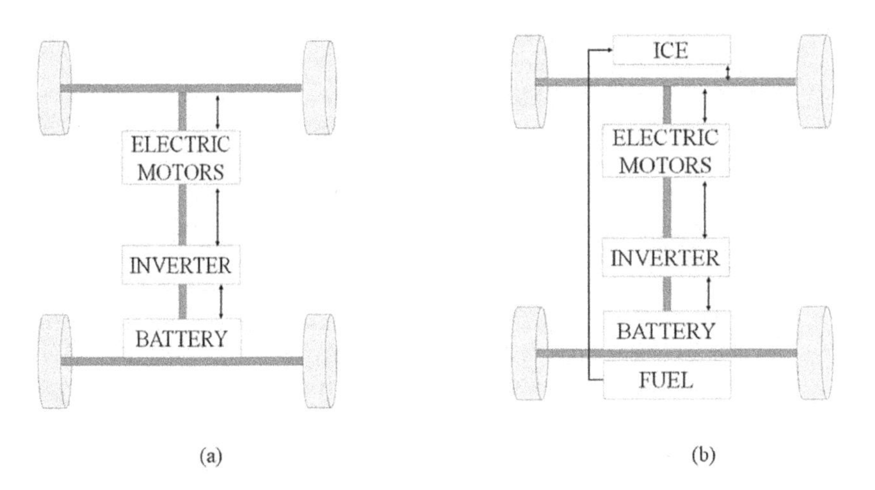

Figure 1.1 Functional block diagram of (a) EV and (b) HEVs.

A. ***DC Motors:*** Although DC motors allow for simple flux and torque control and decoupling, their design (which includes brushes and rings) creates a need for regular maintenance. With the advent of vector control for AC (alternating current) motors, DC motors lost some of their allure in traction applications (synchronous and induction). However, it still has a lot of potential for low-power uses.
B. ***Induction Motors:*** Squirrel cage induction motors have risen to the top of the field because they are dependable, durable, require little upkeep, and can function effectively even in harsh environments. Methods of vector control allow for the separation of torque and field regulation. Increased velocity range can be achieved through the use of flux weakening in the constant power zone. There is a break-down torque in the constant power region, efficiency drops and losses rise with increased speed, induction motors naturally have lower efficiency than permanent magnet motors because of the rotor winding, and inefficient motors have a low power factor. Researchers have taken a number of different approaches to addressing these problems, including using dual inverters to expand the constant power region, incorporating doubly-fed induction motors to achieve high performance at low speeds, and minimising rotor winding losses in the design phase.
C. ***Permanent Magnet Synchronous Motor (PMSM):*** PMSM is the main competitor to induction motors in traction applications. Automakers like Toyota, Honda, and Nissan, among others, have already begun using these drives. The advantages of these motors include increased power density, greater efficiency, and better heat dissipation. However, the constant power region of such motors is necessarily constrained. The speed range and efficiency of PMSM can be expanded by adjusting the conduction angle of the power converter at speeds above the base speed. These motors can lose their magnetism due to heat or an armature reaction, which is a drawback.
D. ***Switched Reluctance Motor:*** These motors' features include fault tolerance, straightforward control, simple and sturdy design, and outstanding torque-speed characteristics. A switching reluctance motor has an inherent ability to function throughout a large constant power zone. For this motor, a number of drawbacks have been mentioned, including excessive noise, high torque ripple, and electromagnetic interference.
E. ***Brushless DC Motors (BLDC):*** In contrast to BLAC motors, which are fed by sinusoidal waves, they are driven by rectangular waves. Their primary benefits include the elimination of brushes, compact design, high efficiency, and high energy density (Table 1.4).

Models introduced into India's automotive market show that PMSM motors are the most popular among other types of drives we've discussed so

Table 1.4 Major EV models in India

Model	*Motor used*
Mercedes-Benz EQS	Two permanently agitated synchronous motors
Tata Tiago EV	PMSM
Tata Nexon EV	PMSM
Kia EV 6	PMSM
Porsche Taycan	PMSM
MG ZS EV	PMSM
Mini Cooper SE	Asynchronous motor
Jaguar I-Pace	EV 400
BMW iX	Three-phase synchronous motor
Hyundai Kona Electric	PMSM
Tata Tigor EV	PMSM
Mahindra E Verito	IM

Source: Cardekho.com.

far. The PMSM has become the standard in the EV industry because of the high torque and power density made possible by high-energy-density PMs (such as neodymium Fe boron (NdFeB) and samarium cobalt (SmCo)) [14]. One can divide the PMSM into two subsets, the internal PMSM and the surface PMSM. The capacity of the internal-overload PMSM is greater for a given size [13]. Therefore, EVs are more likely to have internal PMSMs. High-energy-density PM costs at least twice as much as the sum of the other raw materials used to make electric motors due to its limited yield, non-renewability, and geopolitical nature. Thus, it is crucial and urgent for the EV industry to find methods that can reduce PM costs without significantly reducing performance. The Switched Reluctance motor (SRM) with the cheapest materials is receiving more and more attention. To generate the same 30 kW of output power, the SRM's material cost is less than 80% of that of the IM and roughly half of that of the PMSM (NdFeB) [15]. SRM's drawbacks, which include low torque density, high torque ripple, and high noise, have led to its use in only a small percentage of modern EVs. From research into motor design, it is clear that the PMSM is the most effective EV motor for regenerative braking. Every motor has its unique characteristics which can be utilized with varied requirements of performances [16].

1.5.3.2 Battery

A. ***Ni-MH Batteries:*** The most advanced technology used in hybrid and electric vehicles starting in 2000 was Ni-MH batteries, which were regarded as the forerunners of modern technology.

   Unlike the batteries already in use, particularly Ni-Cd and Lead-Acid batteries, Ni-MH technology was meeting the demands placed on batteries designed to be used in the automotive industry. Batteries with a particular energy density of 70 Wh/kg were able to provide over 300 km of autonomy thanks to their high energy density and power. Also, when used in propulsion systems with electric engines of 320 V AC or 180 V DC, these batteries have a successful longevity (until 80% depth of discharge) [17]. Widespread market acceptability has been facilitated by their capacity to utilise regenerative energy recovered from braking, the use of recyclable materials in their construction, and excellent thermal properties.

B. ***ZEBRA Batteries:*** Some prototype cars and buses intended for urban public transportation were outfitted with Na-$NiCl_2$ batteries, also referred to as Zeolite Battery Research Africa (ZEBRA) batteries. These batteries are outstanding, especially considering their improved energy efficiency and lower price. Increased internal working temperatures (270°C–350°C) are indicative of the large downsizing of the Na-$NiCl_2$ batteries, necessitating constant use of the electric car to prevent the battery electrolyte from freezing [18]. An external heating system that consumes 90 Wh of battery power can be used to maintain the system's operational temperature when the car is not in use.

C. ***Lithium (Li)-Ion Batteries:*** Li-ion batteries are currently the most widely used type of battery technology in EVs because of their high energy density and improved power per mass battery unit. As a result, a number of battery types with lower costs and smaller sizes and weights have been developed. The disadvantage of Li-ion batteries is that they operate at high operating temperatures, which may affect their durability, security, and energy efficiency. In order to regulate and keep track of internal cell temperature, this technology needs a single management battery system. In addition to the drawbacks brought on by the temperature of operation, there are issues with high production costs, the ability to recycle spent batteries, and infrastructure for charging. At present, Li-ion batteries are the most popular market segment for EV components. Because of their low weight, high energy storage capacity, moderate energy consumption (14.7 kWh/100 km), continual cost price decline, advanced production technology, increased cycle life, and moderate energy consumption, Li-ion batteries are the best alternative in this sector [17].

1.5.4 Hybrid vehicles with both IC engine and battery

Parallel or series hybrids are other names for hybrid electric vehicles (HEVs). HEVs have both an engine and an electric motor. The motor is powered by batteries, while the engine is powered by fuel. The transmission is simultaneously turned by the engine and the electric motor. Hence, this propels wheels. HEVs may or may not have a plug-in facility to recharge the batteries. PHEVs mostly run on batteries whereas HEVs don't have a provision to recharge the battery by an external source.

Similar to a regular car, the engine gets its power from the gas tank. Batteries are powered by an electric motor. The engine and electric motor can both turn the transmission at the same time. Regenerative braking will boost a vehicle's mileage even more.

1.5.5 FCEVs (Fuel-cell electric vehicles)

For civilian applications, FCEVs fuelled by hydrogen are still in the early stages of deployment and marketing. Limited numbers of test vehicles are available for particular organisations having access to hydrogen fuelling stations. For instance, roughly 300 fuel cell-powered Mercedes B-Class F-Cells are currently being tested on the roads of Europe and the United States with plans to bring the technology to market in 3 years. Fuel cells convert energy significantly more effective than traditional ICEs, and EVs driven by them emit only water vapour, making them zero-emission automobiles [19].

1.6 FUEL

About 75% of all energy used in the transportation sector in 2006 was for road transportation [8]. This percentage is significantly greater in Europe, where it amounted to 82.5% in 2009 [9]. Light-duty vehicle energy consumption made up the majority of forms of transportation on the road, at roughly 50%–60% [9]. Crude oil is now used nearly exclusively in road transportation. Therefore, to meet GHG emission targets and reduce reliance on oil, the total amount of road vehicle consumption must be substantially reduced.

Although there are many variables that affect GHG intensity, the carbon content of the fuel and the energy consumption of the vehicle have the greatest impact.

1.6.1 Alternative fuel

A. *CNG (Compressed Natural Gas):* Through retrofitting existing gasoline-powered cars, technology for CNG propulsion is mostly employed in taxis and private passenger vehicles. On-board high-pressure gas tanks are used to compress and store natural gas. Engines powered by natural gas utilise the same spark ignition mechanism as engines powered by gasoline. With the addition of fuel injection and ignition systems, ordinary gasoline engines can be converted to run on natural gas. The main benefit of CNG vehicles over traditional gasoline vehicles is their affordable fuel prices. Depending on the cost of CNG and gasoline, a CNG car often has a fuel cost that is 20%–40% less than a gasoline vehicle. Additionally, CNG vehicles have the advantage of lowering oil consumption and GHG emissions. Engine power degradation, limited driving range, less baggage room, and increased maintenance costs are issues that CNG vehicles must deal with.

B. ***Biofuel (Ethanol and Biodiesel):*** To address the demand for transportation fuel, biomass, unlike other renewable energy sources, may be directly transformed into liquid fuels, or "biofuels." The most widely used biofuels today are ethanol and biodiesel, both of which were developed during the first generation of biofuel technology.

The reduction of the country's reliance on foreign energy sources, such as fossil fuels, is a primary objective of energy security efforts. Concern about our reliance on foreign sources to meet our fuel needs, as well as environmental pollution concerns, has prompted the search for alternative fuels that offer superior environmental advantages and are economically competitive with fossil fuels. This foresees biofuels' central role in India's energy source. Agricultural and forestry waste, MSW, cow dung, etc., are all resources that can be used to produce biofuels.

Sugar and starch crops like maize, sugarcane, and sweet sorghum are the primary sources of the carbohydrates used to produce bioethanol. Although ethanol can be used as a car fuel in its pure form (E100), it is most frequently blended with gasoline to raise octane and lower pollutants. One of the actionable points in India's plan to increase the percentage of ethanol blended into gasoline to 10% by 2021–2022 and 20% by 2030 is increasing the country's ethanol distillation capacity.

Through trans-esterification, oils and fats can be converted into biodiesel. It can be used as an automobile fuel in its pure form (B100), although it is more frequently blended with diesel to cut down on hydrocarbon, carbon monoxide, and particulate emissions [20]. India currently produces biodiesel from palm stearin oil that is imported into the country. Used cooking oil (UCO) that is readily available in the domestic market has been prioritised as the feedstock in the effort to eliminate palm stearin and work toward import substitution. The use of UCO as a feedstock for biodiesel development has been explored. For the purpose of conversion, UCO can be gathered from Bulk Consumers like hotels, restaurants, cafeterias, etc.

C. ***Hydrogen as Fuel*** [22]***:*** Hydrogen has a specific energy roughly three times that of gasoline but a volumetric energy only a tenth as great (four times that of gasoline in its liquid condition). In order to reduce the size and weight of the gas storage system, hydrogen must be compressed to 350 or 700 bars. Most light-duty fuel-cell vehicles can travel over 450 km on a single filling of their hydrogen tanks (4–10 kg). When it comes to long-distance driving, fuel-cell systems are the way to go. However, they require hybridisation with batteries and/or super capacitors to deal with the frequent starts and stops, as well as the wildly varying power requirements that come with regular driving. This also improves the system efficiency.

1.6.2 Battery

Nowadays, the most widely utilised battery technology in EVs is Li-ion batteries. As a result, a number of battery types with lower costs and smaller sizes and weights have been developed. Battery packs can have series, parallel, or series–parallel arrangements of Li-ion batteries depending upon the requirements (Table 1.5).

1.7 CONTROL AND COMMUNICATION SYSTEMS

The communication protocols that must be used for communication and control between an EV, a charging station (often referred to as an electrical vehicle supply equipment (EVSE)), and a central management system (CMS) are given below:

1.7.1 Between EVSE and CMS

For the maximum charging rate to be adjusted based on the grid's availability, communication between the CMS at the electrical power utility firm and the EVSE is necessary. This will allow for time-of-use-based metering at various rates. This is necessary because the grid will restrict supplies anytime there is a shortage. Additionally, this will allow customers to reserve chargers. Open charge point protocol will be the communication protocol

Table 1.5 Battery type in major EV models in India

Model	*Battery used*
Mercedes-Benz EQS	Li-ion, 108 kWh. Nickel, cobalt, and manganese are combined in EQS battery cell chemistry in an 8:1:1 ratio, reducing the cobalt content to about 10%
Tata Tiago EV	24 kWh Lithium-ion battery
Tata Nexon EV	30.2 kWh high-energy-density lithium-ion battery
Kia EV 6	77.4 kWh, Nickel-Cobalt-Manganese battery pack

Source: Cardekho.com.

employed. Since open charge point protocol is encrypted, it is somewhat protected from hackers [4].

The standards mainly used are as follows:

- CHAdeMO
- GB/T-27930
- Combined Charging System

For physical communication between the car and the charging station, the combined charging system uses power line carrier communications, in contrast to ChadeMO and GBT, i.e., the Control Area Network (CAN) internal network, which is used for communication with EVSE. Compared to CAN, power line carrier communication links can enable higher data rates.

1.8 CONCLUSION

The transportation system has evolved over decades from animal carts to fast-moving motorised vehicles. Over the period, the research interests in the field have paved the path for even better performing and highly efficient vehicles. Infrastructure development creates an environment to adapt for technological advancements in the field of transportation. The requirement of wheels running on the roads is changing and infrastructure developments must match the pace. Collecting, storing, analysing, using, and disseminating multi-source data is now simpler and less expensive as a result of recent developments in big data and integrated vehicle-infrastructure-pedestrian settings. In addition to increasing the system's versatility, a networked VIP environment enables the introduction of real-time management and control approaches to improve system performance. In a linked environment, people, infrastructure, and vehicles can all communicate with one another utilising a central system that uses a 4G or more recent telecommunication network or a peer-to-peer networking protocol. This is one of the technologies that has the biggest potential to change the urban ecosystem. Information sharing and interactions between vehicles and infrastructure, pedestrians and infrastructure, and pedestrians and vehicles are all feasible.

The battery technology and research studies over the alternative fuel embarked on a new generation of vehicles that can hit the market soon. Na-$NiCl_2$, Ni-MH, Li-ion, and Li-S are used in EVs. Ni-MH and ZEBRA batteries are outnumbered by Li-ion in EV market segment to be the optimal selection. Li-S batteries demonstrated the highest energy consumption, low weight and price than other batteries, and may attract EV manufacturers. Fuel cells using hydrogen as a fuel may emerge as a perspective disruptive technology.

ICE technology has dominated the transportation sector for many decades, but the current market trend prophesied that EVs will see a record growth. Research and development in this field is in their early stages and the century may witness higher efficiency in every sub-system of the EVs with a new era in the transportation system.

REFERENCES

[1] G. J. Van Wylen and R. E. Sonntag, *Fundamentals of Classical Thermodynamics*, 6th ed. John Wiley & Sons, 2002.

[2] "University of Rochester, NY, The growth of the steam engine online history resource, chapter one." History.rochester.edu. Archived from the original on 2012-02-04. Retrieved 2012-01-26.

[3] P. K. Nag, *Power Plant Engineering*. Tata McGraw-Hill. ISBN 0-07-043599-5, 2002.

[4] Report of the Committee on Technical Aspects of Charging Infrastructure for Electric Vehicles, Ministry of Power- Government of India, Central Electricity Authority, March 2018, Available: https://cea.nic.in/wp-content/uploads/2020/04/ev_cea_report.pdf.

[5] A. S. Leyzerovich. "New Benchmarks for Steam Turbine Efficiency. (Steam Turbines)," *Power Engineering*, 106(8), 37–41, 2002.

[6] W. W. Pulkrabek, *Engineering Fundamentals of the Internal Combustion Engine*. Prentice Hall, 1997.

[7] G. Borman and K. Nishiwaki. "Internal-Combustion Engine Heat Transfer," *Progress in Energy and Combustion Science*, 13(1), 1–46, 1987. doi:10.1016/0360-1285(87)90005-0.

[8] International Energy Agency, Transport, Energy & CO_2: Moving Towards Sustainability, OECD/IEA, Paris, France, 2009.

[9] L. Tartakovsky, M. Gutman, & A. Mosyak. Energy Efficiency of Road Vehicles–Trends and Challenges. In Emmanuel F. Santos Cavalcanti and Marcos Ribeiro Barbosa (Ed.), *Energy Efficiency: Methods, Limitations and Challenges*, 63–90. Nova Science Publishers, 2012.

[10] Z. Cao, A. Mahmoudi, S. Kahourzade, and W. L. Soong. "An Overview of Electric Motors for Electric Vehicles," *31st Australasian Universities Power Engineering Conference (AUPEC)*, Perth, Australia, IEEE, 2021. doi:10.1109/AUPEC52110.2021.9597739.

[11] Reitz, H. R. et al., "IJER Editorial: The Future of the Internal Combustion Engine," *International Journal of Engine Research*, 21(1), 3–10, 2019. doi:10.1177/1468087419877990.

[12] N. Hashemnia and B. Asaei, "Comparative Study of Using Different Electric Motors in the Electric Vehicles," *18th International Conference on Electrical Machines*, Vilamoura, Portugal, IEEE, 2009. doi:10.1109/ICELMACH.2008.4800157.

[13] Z. Wang and T. W. Ching, "Challenges Faced by Electric Vehicle Motors and Their Solutions," *IEEE Access*, 9, 5228–5249, 2021. doi:10.1109/ACCESS.2020.3045716.

[14] K. L. V. Iyer, C. Lai, S. Mukundan, H. Dhulipati, K. Mukherjee, and N. C. Kar, "Investigation of Interior Permanent Magnet Motor with Dampers for Electric Vehicle Propulsion and Mitigation of Saliency Effect during Integrated Charging Operation," *IEEE Transactions on Vehicular Technology*, 68(2), 1254–1265, February 2019. doi:10.1109/TVT.2018.2865852.
[15] J. D. Widmer, R. Martin, and M. Kimiabeigi, "Electric Vehicle Traction Motors without Rare Earth Magnets," *Sustainable Materials and Technologies*, 3, 7–13, April 2015. doi:10.1016/j.susmat.2015.02.001.
[16] Z. Yang, F. Shang, I. P. Brown, and M. Krishnamurthy, "Comparative Study of Interior Permanent Magnet, Induction, and Switched Reluctance Motor Drives for EV and HEV Applications," *IEEE Transactions on Transportation Electrification*, 1(3), 245–254, October 2015. doi:10.1109/TTE.2015.2470092.
[17] C. Iclodean, B. Varga, N. Burnete, and D. Cimerdean, "Comparison of Different Battery Types for Electric Vehicles," *IOP Conference Series: Materials Science and Engineering*, 252, 012058, 2017. doi:10.1088/1757-899X/252/1/012058.
[18] T. M. O'Sullivan, C. M. Bingham, and R. E. Clark, "Zebra Battery Technologies for All Electric Smart Car," *International Conference on Power Electronics, Electrical Drives, Automation and Motion, SPEEDAM*, Taormina, Italy, 2006, 1, pp. S34-6-11. doi:10.1109/SPEEDAM.2006.1649778.
[19] C. Arbizzani, F. De Giorgio, and M. Mastragostino, "Battery Parameters for Hybrid Electric Vehicles," In *Advances in Battery Technologies for Electric Vehicles*. Woodhead Publishing, Vol. 1, 55–72, 2015. doi:10.1016/B978-1-78242-377-5.00004-2.
[20] C. M. Bayetero, C. M. Yépez, I. B. Cevallos, and E. H. Rueda, "Effect of the Use of Additives in Biodiesel Blends on the Performance and Opacity of a Diesel Engine," *Materials Today: Proceedings. Advances in Mechanical Engineering Trends*, 49, 93–99, January 2022. doi:10.1016/j.matpr.2021.07.478. ISSN 2214-7853. S2CID 238787289.
[21] S. Kahn Ribeiro et al., "Transport and Its Infrastructure," In Climate Change 2007: Mitigation. Contribution of Working Group III to the Fourth Assessment Report of the Intergovernmental Panel on Climate Change, 2007.
[22] J. Van Mierlo, G. Maggetto, and Ph. Lataire, "Which Energy Source for Road Transport in the Future? A Comparison of Battery, Hybrid and Fuel Cell Vehicles," *Energy Conversion and Management*, 47(7), 2748–2760, 2006. doi:10.1016/j.enconman.2006.02.004.

Chapter 2

Modern transport system

Various categories and transitioning challenges

Rudraksh S. Gupta
Shri Mata Vaishno Devi University

Arjun Tyagi
Netaji Subhas University of Technology

Sanjeev Anand
Shri Mata Vaishno Devi University

CONTENTS

DOI: 10.1201/9781003436089-2

ABBREVIATIONS

BEV Battery Electric Vehicle
HEV Hybrid Electric Vehicle
ICE Internal Combustion Engine
IEC International Electrotechnical Commission
PHEV Plug-in Hybrid Electric Vehicle
SAE Society of Automotive Engineers

2.1 INTRODUCTION

Mobility does not indicate simply commuting from one location to another. Its worth rests in the convenience it offers and how this enhances the functionality and standard of living for both individuals and society as a whole. Similar to this, the transportation system encompasses much more than different vehicles that use our roads, bridges, railroads, ports, and airports. Instead, this system facilitates commercial operations for organisations and corporations and has an impact on how individuals undertake their lives. The effective migration of people and things, which is the foundation of our economy, depends on mobility. Mobility expanded and changed employment opportunities throughout the 20th century, creating new markets and influencing how the land was used. All of this resulted in long-term gains for local economic development. Throughout reality, consumer demand for products and services had a big impact on how people moved about in the 20th century in terms of emigration, employment, and transportation. Social cohesion, opportunity expansion, and health and well-being improvements all depend on mobility. It has altered society, the layout, and the placement of towns and services and has significantly influenced personal decision-making. However, there are still disparities in people's mobility. There are ongoing problems with some people's lack of mobility and the strain it throws on others, such as those who must spend a large percentage of their income on transportation.

Mobility and transportation have been through a period of upheaval over the last several decades, which are far from over. In the late 19th century, when bicycles first became widely available, a reliable public transportation system was finally put in place. The advent and fast expansion of four-wheeled vehicles coincided with the rise of modern infrastructure, increased business activity, and higher levels of personal discretionary money. Since the first mass-produced car was made in 1896, the automotive industry has seen fast expansion and profound changes. The growing and extremely fluctuating price of oil presents a challenge to the economy and society at a time when demand for mobility, traffic volume, and the associated requirement for transportation infrastructure are all on the rise. Policymakers and planners are trying to balance multiple priorities at once: they must combat climate change by lowering CO_2 emissions and energy use, and they must deal with the fallout from the financial crisis, which has left them with far less money to invest in national infrastructure maintenance and development.

However, the current transportation sector is anticipating yet another shift to electric variants in practically every feasible domain of the transport industry, such as road, water, and air, owing to many difficulties, such as global warming, finite fossil fuel, and climate change challenges. The advantages of electric vehicles (EVs) have come to the fore in recent years as concerns about oil resource depletion and environmental pollution have gained greater media attention. One strategy for developing a sustainable and environmentally friendly transportation system that has emerged in recent years is the electrification of transportation networks. Incentives, research programmes, and operational and infrastructural testing are some of the ways in which many nations contribute to the development of an electric mobility plan.

To propel a vehicle, electricity is a viable alternative to oil. Therefore, it guarantees a reliable supply of power for the transportation industry. There is potential for this industry to make use of a vast array of alternative renewable and carbon-free energy sources to aid national efforts to lower greenhouse gas emissions. While modifications are being made to all modes of transportation—including road, marine and air—the road system is receiving the most attention since it is the most prevalent mode of transportation and has the most well-developed infrastructure. Multiple countries throughout the globe have approved the transition to EVs since 2015, after the signing of the Paris accord. As can be observed in Figure 2.1, the rate of adoption has been steadily increasing since 2015. The future adoption rate under the sustainable development scenario can be observed in Figure 2.2 as collected from IEA annual report [1].

2.2 CLASSIFICATION OF EVs

There are primarily three categories of electric cars that have evolved from the prototype stage to the large-scale production phase of development at present. They consist of hybrid electric vehicles (HEVs), plug-in hybrid electric vehicles (PHEVs), and battery electric vehicles (BEVs). The different

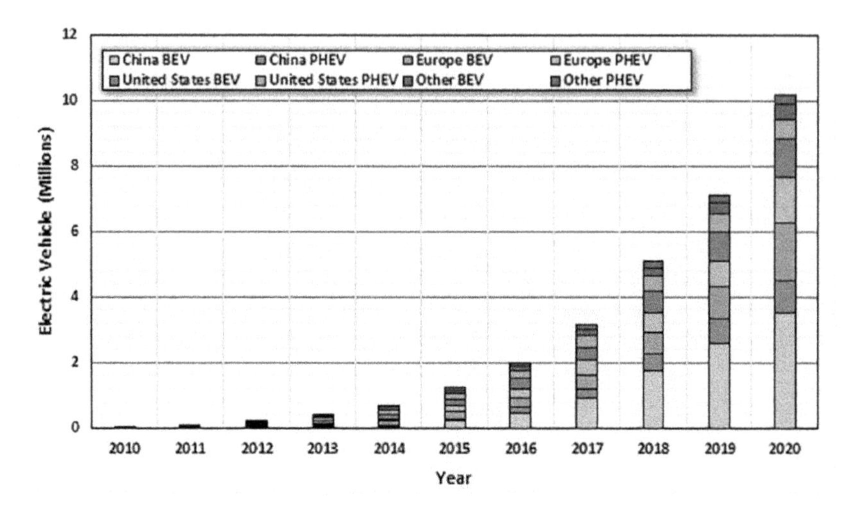

Figure 2.1 Electric vehicle stock Y-o-Y basis [1].

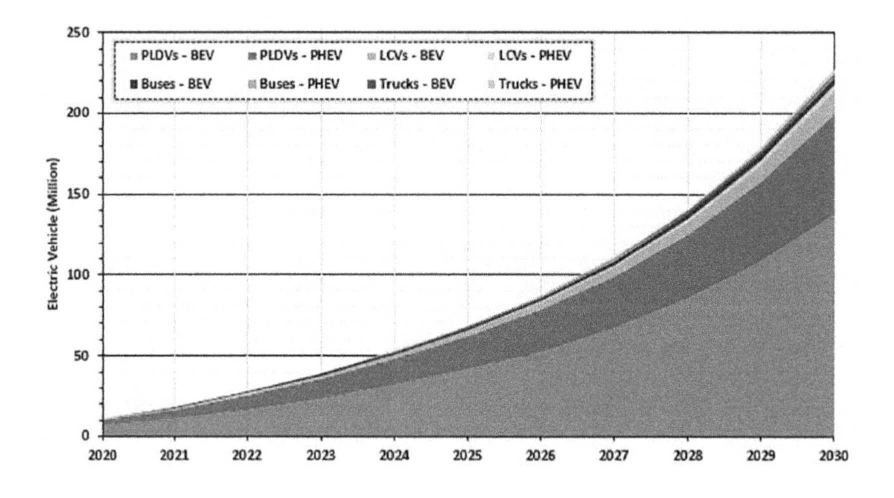

Figure 2.2 Electric vehicle stock forecast under sustainable development scenario [1].

types of EVs are shown in Figure 2.3. The basic comparison is shown in tabular form in Table 2.1 for a different EV, and Table 2.2 provides advantages and disadvantages for a different EV.

2.2.1 Hybrid electric vehicles

This type of vehicle is a subset of hybrid automobiles that get their propulsion power from a combination of fuel and electrical power. There are

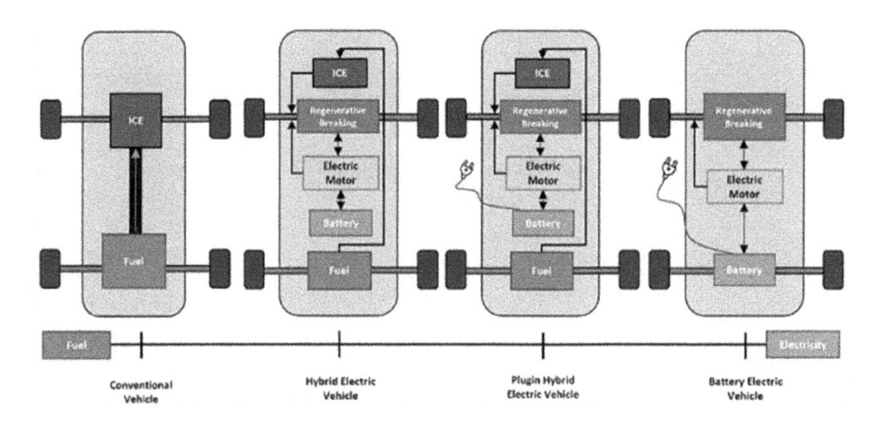

Figure 2.3 Classification of electric vehicles [2, 3].

Table 2.1 Comparison of different electric vehicles [14]

Vehicle type	*Internal combustion engine*	*Battery charging*	*Driving range*	*Top speed*	*Electricity consumption (kWh/km)*	*CO_2 emissions (gCO^2/km)*
Hybrid electric vehicles	YES	On-board (internal)	900–1200 (hybrid)	-	NA	109
Plug-in hybrid electric vehicles	YES	On-board (internal) and external charging	900 (hybrid)	160	0.225	132
Battery electric vehicles	NO	External charging	120–600	80–200	0.175	88

two types of propulsion systems: an internal combustion engine (ICE) and a battery-operated electric motor (EM) setup. Better fuel efficiency or increased performance over a normal ICE car is the goal of having an on-board battery and EM combination. Several different varieties of HEVs, each with its own unique combination of the ICE and EM, may be found on the market today.

The size of the ICE and the EM may have a major impact on the HEV's control approach. The hybridisation ratio refers to the power output of the EM as a percentage of the total power output of the power train. In a vehicle with a high hybridisation ratio, the electric route (EM and battery) is enlarged, while the ICE is downsized. While this is true, a low hybridisation ratio actually causes a shorter electric route and larger ICE. Due to the motor/generator system's capacity to serve as a generator, the HEV's battery may be charged whether the vehicle is in motion or stopped.

Table 2.2 Advantages and disadvantages of a different electric vehicle

S. No.	*Electric vehicle type*	*Advantages*	*Disadvantages*
1	HEV	• Reduced fossil fuel usage. • Regenerative braking system.	• Less electric range. • No on-board charging. • High maintenance cost. • Complex system. • Similar impact on environment as of ice vehicle.
2	PHEV	• On-board charging. • Larger battery capacity than HEV. • Little carbon emission. • Regenerative braking system.	• Complex system. • High maintenance. • Overall less electric range.
3	BEV	• Most environmentally friendly. • Largest electric driving range. • Zero carbon emission. • Zero dependence on oil.	• Overall lower range. • High charging time. • Battery-dependent range.

In both instances, the ICE will be managed by the central processing unit of the vehicle to run the generator and charge the battery. Thus, regenerative breaking technology will be used to charge the battery. In this system, the deceleration energy from a vehicle's brakes is absorbed and stored, enabling a steady flow of power independent of the direction the vehicle is moving. The EM is put into reverse by the regenerative brakes to slow the vehicle, which makes the wheels spin the other way. When the automobile is in reverse, the power produced by the engine is utilised to replenish the batteries. The efficiency of regenerative brakes peaks during slower speeds and during stop-and-go driving conditions, such as in city traffic. Friction brakes are also employed in hybrid cars as a backup to regenerative braking technology. In order to enhance the range of HEVs, nickel-cadmium (NiMH) battery packs are most commonly used and are routinely charged up to 40%–60% of their full capacity [4–7].

2.2.2 Plug-in hybrid electric vehicles

PHEVs are an innovative and promising field of transportation research and development. They operate similarly to HEVs in that they rely on both an ICE and a battery pack for propulsion. Specifically, PHEVs are HEVs with a battery bank of 4 kWh or more, the capacity to recharge the battery from an external source, and a travel distance of at least 16 km in purely electric mode [8]. Using either electricity, fossil fuel, or a combination of the two to propel a vehicle is possible. These vehicles' wide variety of potential advantages, including reduced dependency on oil, increased fuel economy, higher power efficiency, lowered greenhouse gas emissions, and the advantages of vehicle-to-grid (V2G) technology, have led to their increasing popularity in recent years. PHEVs have the potential to outperform HEVs in terms of efficiency in terms of mileage since a better restricted and controlled ICE usage would tend to boost overall combined vehicle efficiency and enable the ICE to be employed even closer to its peak efficiency by functioning only at high vehicle speeds [9,10].

The PHEV industry currently has three distinct designs. Instead of using the ICE for propulsion, as is the case with parallel designs, the EM acts as the only source of propulsion in a series design. In a parallel configuration, propulsion may come from either an ICE or an EM. The third form, a series hybrid, may function in either a parallel or a series configuration flexibility [9, 11]. For PHEVs as a whole, two distinct battery technologies are viable options. The lithium-ion (Li-ion) and nickel-cadmium (NiMH) batteries fall within this category. NiMH batteries have lower energy and power densities than Li-ion batteries, resulting in a reduced EV range, slower top speeds, and slower acceleration times. NiMH batteries, on the other hand, have the potential to be more long-lasting than Li-ion ones and to withstand a greater number of deep-discharge cycles, up to 80% [12].

2.2.3 Battery electric vehicles

BEVs do not use an ICE of any kind for propulsion. Rechargeable battery packs, capacitors, and flywheels may all be used to produce electricity on board. Like PHEVs, the battery may be charged using either a conventional household electrical socket or a special public charging station. When compared to conventional ICE automobiles, BEVs have the potential to dramatically scale back the harmful greenhouse gases produced by the transportation industry, much like the possibilities offered by HEVs and PHEVs. Even while PHEVs have the ability to reduce emissions, BEVs have a considerably larger potential for doing so. BEVs have advantages in performance over conventional petrol cars in addition to the potential environmental benefits.

The installation of high-capacity battery packs included in is responsible for these advantages. These battery packs are responsible for providing electric power to motors with intrinsically higher torque at reduced vehicle speeds compared to ICE, enabling BEVs to be much more rapid and accelerate from a halt quicker than conventional automobiles without the need for gears or clutches. However, the lack of an ICE means that the BEV's ability to heat the cabin is significantly diminished, which is an often-overlooked disadvantage. In colder climates, this may become a serious problem, making it all the more urgent to find a remedy.

A BEV's fundamental systems consist of a motor controller, EM, and a battery pack. In a steady state, the battery pack powers the controller, which regulates and controls the power being made available to the EM for the propulsion of vehicle. The location of the accelerator pedal determines the amount of power sent to the motor by the controller. Since there is no ICE in a BEV, the battery must be used at all times, unlike a PHEV. Additionally, the battery energy capacity has to be strong enough to ensure at least a certain driving efficiency that would be adequate to cover a daily regular driving cycle for BEV drivers to feel comfortable with their purchase. As a result, lightweight, high-power battery packs are required, ideally ones that can resist heavy loads while both charging and discharging. However, the charging time would increase dramatically with a higher energy density battery. NiMH or Li-ion batteries are the only ones that can provide such functionality, but they are prohibitively expensive. The high price of batteries is currently the biggest factor in the overall price of BEVs and the major barrier to the widespread commercial viability of BEVs [2, 3, 13]. Table 2.1 provides a comparison between different EVs under various variables.

2.3 CHARGING INFRASTRUCTURE

Charging infrastructure acts as a parallax to fuel pumps for conventional vehicles. This section gives an overview of the various charging levels, charging modes, and charging schemes for EVs, as well as a brief discussion of a few international standards that should be considered throughout the Electric Vehicle Charging Station (EVCS) implementation process.

2.3.1 Charging based on different levels

The EVCS functions like a gas station in that it supplies electric power for recharging EVs. The user interface panel, charging wire, and charging port all function as a single charging hub. Electric outlets are wired in a certain way based on the grid's specifications for voltage, frequency, and transmission standards. Important roles in classifying charging levels and modes between countries, along with varying safety regulations, are played by the Electric Power Research Institute, the Society of Automotive Engineers (SAE), and the International Electrotechnical Commission (IEC). Alternating current (AC) Level 1, AC Level 2, and AC Level 3, also known as Direct Current Fast Charging (DCFC), are the various degrees of charging according to established standards. In accordance with the revised rules, there are now two distinct categories of Direct Current Charging (DCC) [1]: Level-1 DC and Level-2 DC. Level-1 AC and Level-2 AC are the most common charging standards due to a preference for residential charging. Figure 2.4 represents the different charging levels and their utility zone [2, 15].

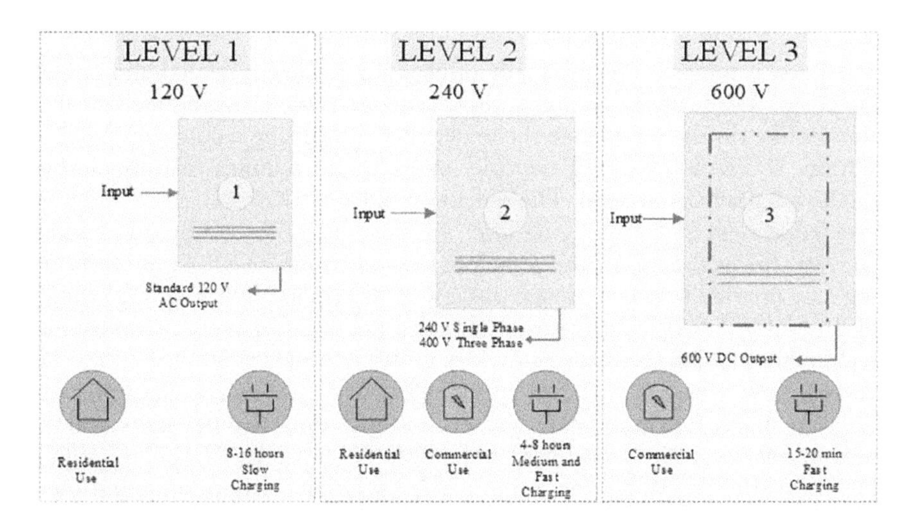

Figure 2.4 Different charging levels.

2.3.1.1 Level-1 charging

For Level-1 charging, a regular 120 V wall outlet is utilised. These chargers can only handle a maximum of 16 A of power. A complete charge may take anywhere from 8 to 16 hours, depending on the size of the battery when utilising a 1.9 kW charging connection. A SAE J1772 standard connection is utilised to connect the EV to the charging pillar. Level-1 charging is the most affordable in nature but also the slowest in terms of charging. Level-1 charging when implied with a tariff-based charging scheme can be used to further reduce the charging cost by discharging it during peak load time via V2G technology and charging during low-price off-peak time via G2V technology.

2.3.1.2 Level-2 charging

Level-2 charging stations are most often and prominently utilised for charging at both home and public charging facilities. For private and domestic installations, a single phase 240 V, 40 A current supply is an acceptable standard, whereas a three-phase 400 V AC, 80 A current supply on a full load capacity is required for public charging stations. Level-2 charging systems may offer up to 19.2 kW of electrical power and charge an EV in 4–8 hours. The Level-2 charging approach is widely utilised since it uses the same electric circuits and standard voltage rating as household appliances, notably 240 V. Fast-charging capability adds another major factor provided by Level-2 charging. In terms of protection, both overcurrent voltage and overcurrent protection are included in the Level-2 charging system [2].

2.3.1.3 *Level-3 charging*

The Level 3 substation is primarily utilised for both commercial and public purposes. The primary goal of a direct current (DC) charging facility at a station like this is to replicate the experience of utilising a conventional gas station. DC rapid charging typically takes 15–20 minutes to fully charge (80%) a battery from dead. The remaining 20% is always charged at a low rate, regardless of the charging level. In most cases, the off-board charger converts the incoming AC to DC before feeding it into the EV. Level-3 charging may provide between 36 and 240 kW of electricity, with a voltage range of 200–600 V. SAE categorises DCFC as Level-1 and Level-2. Comparing DC Levels 1 and 2, the former provides 36 kW of power at 80 A of allowable current flow, whereas the latter provides 90 kW at 200 A. DC charging stations are most often found in public settings, including airports, gas stations, movie theatres, and government buildings. The SAE J1772/IEC 62,196-3 DC charger connection is the recommended standard by both organisations. The high cost of installation (about 250,000 INR) is the biggest hindrance to the widespread adoption of DC fast-charging stations [2].

2.3.2 Charging based on charge transfer pattern

Charging a battery may include a number of methods, all of which involve the regulation of current flow. In order to charge an EV's battery, the technology uses rectifiers to change the current from alternating to direct. Multiple methods, including conductive charging, inductive charging, and battery switching, contribute to this charge transfer.

2.3.2.1 *Inductive/wireless charging*

Wireless charging/inductive charging uses an electromagnetic field to transmit electricity from a charging station to an EV. In this method, there is no physical connection between the EV and the power grid in any way. The main drawback of this approach is that it is quite expensive than other approaches and is not as efficient as conductive charging. However, wireless technology has been shown to be 86% effective at charging stations [15, 16] The primary responsibility still lies in the range at which it can transmit charge for recharging EVs. Inductive Power Transfer, Permanent Magnet Coupled transfer, Coupled Magnetic Resonance, laser, and microwave or radio-wave transmission are all examples of subtypes of wireless power. Figure 2.5 represents the wireless charging technique [16].

2.3.2.2 *Conductive charging*

The principle of conductive charging is using a cable to make direct contact between an EV and a charging outlet to transfer power. The least amount of energy is lost during the transmission of power while conducting charging.

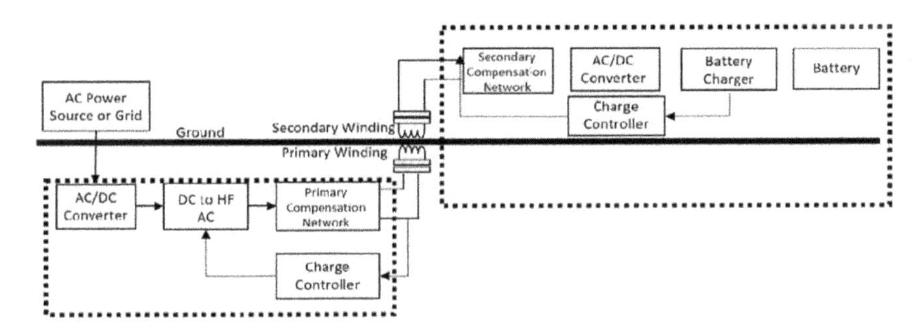

Figure 2.5 Wireless charging system.

Numerous benefits of conductive charging include its economic viability, ability to allow quick charging, ease of use, and high level of efficiency. The on-board system, where complete charging activities like AC–DC conversion occur within an EV, is often a subdued charger. The off-board method, however, offers quick charging. Off-board charging facilities lower EV weight further, allowing for a few kilometre increases in range.

2.3.2.3 Battery swapping

With the battery switching procedure, a fresh, recharged battery may be swapped out for the depleted one at battery swap stations. This method offers a number of advantages, including a significant decrease in refuelling time, low management costs, and reduced charging burden on the grid all through peak hours. However, various drawbacks, including a high upfront cost, huge manpower, and a huge amount of area, are obstacles to the adoption of such procedures.

2.4 HEADWINDS FOR TRANSFORMATION IN THE TRANSPORT SECTOR

2.4.1 World population growth

The global population will continue to rise, increasing the demand for products, environmental assets, and energy. Consumption in developing nations is increasing in tandem with a redistribution of wealth. The increased population may result in inefficient transportation infrastructure, whereas urbanised regions must deal with increased traffic loads. Therefore, a call for transformation in transformation is required and can't be neglected [17].

2.4.2 Demographic and social change

Demographic and social development around the globe will be reflected in the ever-increasing proportion of the population over 65 in the future.

The demographic ageing of the population will have a substantial influence on mobility demands and services. Adults now travel more than earlier generations because they are healthier, richer, and live a more active lifestyle. Social changes such as shrinking family sizes impact housing demand, resulting in new settlement layouts and traffic movements [17].

2.4.3 Globalisation 2.0

Globalisation 2.0 refers to a qualitative shift in global economic development, which will continue to be significant in the next decades, with developing economies witnessing stronger rates of growth. The economic boom in the BRIC (Brazil, Russia, India, China) nations is causing a worldwide change in political and financial power, opulence, and economic specialisation, as well as a "multi-polarisation" of the globe with international regulatory gaps, such as those relating to the internet. As wealth is redistributed around the world, there will be repercussions for the consumption of mobility services, the acquisition of vehicles, and the creation of transportation technologies that must take into account the requirements of people from a wider range of cultural backgrounds than in previous decades [17].

2.4.4 Urbanisation

The process of urbanisation, or the movement of people and goods to major cities, is accelerating in developing countries, but it is also anticipated to continue, although at a slower pace. Transportation efficiency is improving as more people and job markets move towards urban cores. Congestion occurs when there is a sudden increase in the number of vehicles on the road. Because of these divergent tendencies, the transportation industry's bottom line, its physical infrastructure, and its daily operations are being pulled in different directions. One of the next decades' greatest problems will be making transportation available to all people at a realistic cost. These new transport reforms need to be built in to cater for the needs [14, 17].

2.4.5 Climate change issues

Global pressure issues such as climate change, environmental degradation, and environmental ethics provide substantial negative externalities to society. To overcome the challenges and cover the costs of climate change, new policies and initiatives are required, which will be incurred via damage and the need for appropriate adaptation measures. The governmental actions used to protect the environment, such as taxes and laws to reduce emissions, are driving technological progress. Awareness of environmental concerns among humans is growing. A shift in consumer behaviour towards more environmentally friendly means of transportation is possible if the market responds to this trend by producing complementary goods and services [14, 17].

2.5 TAILWINDS IN TRANSFORMATION IN THE TRANSPORT SECTOR

2.5.1 Range

The range of an EV is the maximum distance it can drive on a single charge. The range is greatly dependent on the capacity and functioning of the battery. Battery life may decrease with age. Hence, the range can be equivalent to a regular ICE car or even throughout its life phases. How far a car can go is also affected by the way its operator drives and how often the vehicle's auxiliary systems are used. As noted, before, this reliance might deliver varying values across masses. It is possible that few feel it adequate, while the majority will not. It should be noted that those who live in more remote locations tend to see the range as a greater issue. Most EVs have a range of fewer than 350 km on a single charge, compared to the same model powered by conventional fuel [14, 16, 18–20].

2.5.2 Technical immaturity

The general public views this technology as still being in its infancy; hence, they are hesitant to use it until it has been refined. More research and development are needed before the technology can be considered a credible alternative to traditional ICE systems. There aren't any other chassis configurations or powertrain setups to choose from (like sedans, hatchbacks, and pickup trucks). In general, the consensus technological gap is considered a significant obstacle since our current state of knowledge cannot provide the same practical benefits as a regular ICE car [14, 16, 18–20].

2.5.3 Cost of ownership

Pricewise, EVs are above CVs. Reasons for the high cost include the battery's high capital costs and large size, the priority placed on innovative technology, and the lack of economies of scale. Moreover, the overall purchase cost of EVs is significantly greater, leading to a longer prolonged payback time due to expensive raw material prices, higher operating expenses, and insufficient infrastructure. In many countries, the price of an EV is still much more than the price of a CV, despite subsidies and other incentives. Despite significant fuel savings, consumers remain hesitant to pay substantial EV prices. After considering available options, their financial constraints, and their highest priorities, they choose the one that maximises utility [14, 16, 18–20].

2.5.4 Infrastructural challenges

A vehicle's ability to be plugged in is enabled by charging infrastructure, which encompasses a coupler, battery charger, connectors, attachment plugs, and all other requisite equipment, as well as the receptacle device that connects the attachment connector to electrical lines, transformers, and other devices. It is the backbone of introducing EVs to the people in cities. Public charging stations are few and often positioned in awkward places in relation to customers' homes to avoid overloading problems. A good charging network is required for EVs to become mainstream modes of transportation. Due to limited infrastructure, the range anxiety is elevated and thus causes hindrance in the adoption of EVs [14, 16, 18–20].

2.5.5 Impact on electric grid

The introduction of EVs would put additional stress on the existing system. Public opinion may be influenced in a positive direction, which might accelerate the rate of EV adoption, but only if the power and distribution infrastructures are managed efficiently. However, EVs pose a new problem for the existing infrastructure. Public charging stations are still in the process of being installed in Indian cities, and home charging is made more challenging by the country's antiquated electrical system. The present infrastructure is overstretched, and there is a severe lack of charging stations. The uncertainty associated with the current electric grid network is opposed to the spread of EVs [14, 16, 18–20].

2.5.6 Safety

Among the difficulties that EV owners must overcome is worried about public security and reliability. Energy storage in batteries presents more safety issues, such as explosions from overcharging and overheating and hence must be avoided. Lithium batteries and other modern advances in battery technology increase energy density but also increase the potential for malfunction. This occurs when a cell's temperature rises as a result of an electrical current, and those neighbouring cells experience a similar rise in temperature. It follows that EVs are thought to be weather-sensitive owing to their on-board batteries. Batteries' natural deterioration with time is another potential threat to their users' security. Dendrites are a kind of fibre formed on the carbon anodes of ageing batteries by microscopic lithium particles [14, 16, 18–20].

2.6 IMPACT OF EV ON DISTRIBUTION NETWORK AND ELECTRICITY MARKET

While the increased sales of EVs are not expected to significantly raise overall power consumption for time being, they are expected to alter the form of the energy load curve. Most noticeably, evening peak loads will rise as more people charge their EVs after a long day at work or completing day errands. However, the shifting load curve will provide regional difficulties, since the distribution of EVs is expected to vary, sometimes dramatically. There will be significant gains in local peak loads in these residential hot spots and other concentration locations of EV charging, such as public EV-fast-charging stations and commercial-vehicle depots.

One of the significant negative consequences of EV charging stations is the risk they pose to voltage stability and frequency regulation. Voltage stability may be conceptualised as the power system's propensity to maintain constant voltages across all buses after the removal of disturbances. One of the main causes of voltage instability is an unexpected rise in load. Charging EVs causes a spike in demand, which may cause voltage fluctuations. According to studies, in order to keep the grid stable and prevent overloading, 93% of EV demand must be moved to off-peak load hours. On the other hand, 30% EV adoption was shown to raise peak power demands by 53% [21].

However, not all the effects from EVs can be considered hazardous. With incorporation of EV, V2G technology can be used to provide a backup to electric grid during overloading condition of grid and G2V technology to charge the batteries at a lower rate and store energy from solar, wind, and other sources during times of excess availability. The major advantage of V2G technology constitutes peak load shaving, load management, line loss reduction, stabilising grid voltage, and frequency control. With this technology, people can even earn some money where they can charge the EV during odd load period and discharge during peak hours at higher rates.

2.7 CONCLUSION

The modern transport system is need of the hour to limit climate change and global warming. In this chapter, different types of EVs such as HEV, PHEV, and BEV are discussed with availability of different charging levels such as Level 1, 2, 3 for charging EVs. The advantages and disadvantages of each vehicle type and charging levels have been discussed over each other to provide a better understanding while purchasing EVs. EV being a technology in a nascent stage of development shows out few tails' winds in adoption of technology which have been discussed. Some headwinds that boost the adoption of EV have also been discussed with climate change seen as a major factor. To make a redundant system, it is necessary that all the governments around the world in conjunction with private sector must endeavour in the direction of improving the technology of modern transport systems and expand the infrastructure required to sustain the transport network. Achieving the sustainable development goal and Paris accord will require navigating the headwinds and tailwinds of adopting electric transportation systems.

REFERENCES

[1] "Global EV Outlook 2021 – Analysis - IEA." https://www.iea.org/reports/global-ev-outlook-2021 (accessed Sep. 06, 2022).

[2] R. S. Gupta, A. Tyagi, and S. Anand, "Optimal allocation of electric vehicles charging infrastructure, policies and future trends," *J. Energy Storage*, vol. 43, p. 103291, Nov. 2021, doi: 10.1016/J.EST.2021.103291.

[3] E. Fantin Irudaya Raj and M. Appadurai, "The hybrid electric vehicle (HEV)—An overview," *Emerging Solutions for e-Mobility and Smart Grids: Select Proceedings of ICRES 2020*, pp. 25–36, 2021, doi: 10.1007/978-981-16-0719-6_3.

[4] Y. Wang, A. Biswas, R. Rodriguez, Z. Keshavarz-Motamed, and A. Emadi, "Hybrid electric vehicle specific engines: State-of-the-art review," *Energy Rep.*, vol. 8, pp. 832–851, Nov. 2022, doi: 10.1016/J.EGYR.2021.11.265.

[5] F. Mocera and A. Somà, "A review of hybrid electric architectures in construction, handling and agriculture machines," *New Perspect. Electr. Veh.*, Mar. 2022, doi: 10.5772/INTECHOPEN.99132.

[6] X. Zhang, C. C. Mi, and C. Yin, "Active-charging based powertrain control in series hybrid electric vehicles for efficiency improvement and battery lifetime extension," *J. Power Sources*, vol. 245, pp. 292–300, Jan. 2014, doi: 10.1016/J.JPOWSOUR.2013.06.117.

[7] S. Han, S. Han, and H. Aki, "A practical battery wear model for electric vehicle charging applications," *Appl. Energy*, vol. 113, pp. 1100–1108, Jan. 2014, doi: 10.1016/J.APENERGY.2013.08.062.

[8] B. G. Pollet, I. Staffell, and J. L. Shang, "Current status of hybrid, battery and fuel cell electric vehicles: From electrochemistry to market prospects," *Electrochim. Acta*, vol. 84, pp. 235–249, Dec. 2012, doi: 10.1016/J.ELECTACTA.2012.03.172.

[9] S. Amjad, S. Neelakrishnan, and R. Rudramoorthy, "Review of design considerations and technological challenges for successful development and deployment of plug-in hybrid electric vehicles," *Renew. Sustain. Energy Rev.*, vol. 14, no. 3, pp. 1104–1110, Apr. 2010, doi: 10.1016/J.RSER.2009.11.001.

[10] B. Soares M.C. Borba, A. Szklo, and R. Schaeffer, "Plug-in hybrid electric vehicles as a way to maximize the integration of variable renewable energy in power systems: The case of wind generation in northeastern Brazil," *Energy*, vol. 37, no. 1, pp. 469–481, Jan. 2012, doi: 10.1016/J.ENERGY.2011.11.008.

[11] L. P. Zhang, W. Liu, and B. Qi, "Innovation design and optimization management of a new drive system for plug-in hybrid electric vehicles," *Energy*, vol. 186, p. 115823, Nov. 2019, doi: 10.1016/J.ENERGY.2019.07.153.

[12] H. Park, "A design of air flow configuration for cooling lithium ion battery in hybrid electric vehicles," *J. Power Sources*, vol. 239, pp. 30–36, Oct. 2013, doi: 10.1016/J.JPOWSOUR.2013.03.102.

[13] S. Soylu, *Electric Vehicles: The Benefits and Barriers*. InTech, 2010. [Online]. Available: https://books.google.co.in/books?id=VfeZDwAAQBAJ&printsec=copyright&redir_esc=y#v=onepage&q&f=false

[14] A. Poullikkas, "Sustainable options for electric vehicle technologies," *Renew. Sustain. Energy Rev.*, vol. 41, pp. 1277–1287, 2015, doi: 10.1016/J.RSER.2014.09.016.

[15] B. Zhou, G. Chen, T. Huang, Q. Song, and Y. Yuan, "Planning PEV fast-charging stations using data-driven distributionally robust optimization approach based on Φ-divergence," *IEEE Trans. Transp. Electrif.*, vol. 7782, no. c, pp. 1–10, 2020, doi: 10.1109/TTE.2020.2971825.

[16] R. S. Gupta, A. Tyagi, V. V. Tyagi, Y. Anand, A. Sawhney, and S. Anand, "Renewable energy-driven charging station for electric vehicles," vol. 43, pp. 57–78, 2021, doi: 10.1007/978-981-16-1256-5_5.

[17] M. Hoppe et al., "Transformation in transportation?," *Eur. J Futures Res.*, vol. 2, p. 45, 2014, doi: 10.1007/s40309-014-0045-6.

[18] R. Chhikara, R. Garg, S. Chhabra, U. Karnatak, and G. Agrawal, "Factors affecting adoption of electric vehicles in India: An exploratory study," *Transp. Res. Part D Transp. Environ.*, vol. 100, p. 103084, Nov. 2021, doi: 10.1016/J.TRD.2021.103084.

[19] V. Singh Patyal, R. Kumar, and S. Singh Kushwah, "Modeling barriers to the adoption of electric vehicles: An Indian perspective," *Energy*, vol. 237, p. 121554, Dec. 2021, doi: 10.1016/J.ENERGY.2021.121554.

[20] P. K. Tarei, P. Chand, and H. Gupta, "Barriers to the adoption of electric vehicles: Evidence from India," *J. Clean. Prod.*, vol. 291, Apr. 2021, doi: 10.1016/J.JCLEPRO.2021.125847.

[21] Z. Wang and R. Paranjape, "An evaluation of electric vehicle penetration under demand response in a multi-agent based simulation," *Proc. -2014 Electr. Power Energy Conf. EPEC 2014*, pp. 220–225, Feb. 2014, doi: 10.1109/EPEC.2014.14.

Chapter 3

Intelligent control for PMSM drive train

Shilpa Ranjan, Monika Verma, Madhusudan Singh, and Mini Sreejeth
Delhi Technological University

CONTENTS

3.1 INTRODUCTION

Over the past few decades, intelligent control systems (ICS) have quickly replaced conventional problem-solving methods in many fields of technology and science. Power networks, aircraft, autonomous vehicles, and communication networks are some examples of man-made engineering systems that are becoming more complex because of their intricate mathematical expressions. As a result, it is getting more difficult to understand their behavior and generate precise dynamical models to explain them, which inhibits researcher's ability to manage them and regulate their behavior [1, 2]. Intelligent control is an advancement of conventional control, which is used to monitor complex systems that are inaccessible by the existing technique. Intelligent control refers to methods, used for designing, modeling, identifying, and operating control systems that incorporate use of artificial intelligence tools. Such tools include fuzzy logic (FL), neural networks (NNs), machine learning, evolutionary computation, and

DOI: 10.1201/9781003436089-3

genetic algorithms [3]. Since many control issues today cannot be described mathematically or analyzed using the 'conventional control' methods, which were developed earlier to govern dynamical systems, ICS are much needed. The major benefit of an intelligent controller is that it does not require an exact mathematical analysis of the system [4]. In addition to this benefit, the ICS reduce system design time and provide aid in resolving response settling time issues, caused by the employment of mathematical modeling. This realization has been the major motivation behind the proposed study of using intelligent control in an electric drive system (EDS).

Traditionally, the proportional-integral (PI) controller and other adaptive controllers, such as model reference adaptive [5], sliding mode [6], and variable structure controllers [7], are used to resolve the challenges that are faced during the control of EDS. Exact machine models and precise model parameters are required for these controller designs. It is always quite challenging to obtain exact precise parameters of EDS [8]. Generally, high-performance EDS, utilized in the automotive industries, demands for rapid and accurate responses [9]. The permanent magnet synchronous motor (PMSM) is widely used for variable-speed drive systems because of its high torque–current ratio, large power–weight ratio, high efficiency, high power factor, and stability [10]. But it is difficult to compute exact values of d-q axis reactance parameters of PMSM, while implementing any control technique, for obtaining efficient PMSM-based drive systems. It requires a complex design process for implementation of traditional controllers. However, the nonlinear coupling between a drive's winding currents and rotor speed, as well as the nonlinearity present in the electromagnetic torque owing to magnetic saturation of the rotor core, complicates the accurate speed control of these drives [11, 12].

In addition to the benefits like fast and accurate response, the PMSM-based drives also have some drawbacks. Apart from high cost, the sluggishness also arises in the system. It occurs due to the presence of high torque ripples, observed at low-speed applications. Torque ripples may cause speed fluctuations with high noise and vibration [13]. Many control strategies have been developed to help in minimizing the aforementioned drawbacks [14, 15]. There have been a number of speed-based control approaches, where the torque ripple is generated through the speed feedback. In general, either machine design or control designs can reduce the torque ripples produced in the PMSMs [16]. To reduce torque ripples, a number of machine designs are presented, such as by using a smaller number of slots per pole, skewing magnets or stator lamination slots, and reshaping magnets. However, specialized machine designs make the production process more complex, which increases the price of the machine [17]. The torque ripples can also be reduced by controlling the system with PI [18], deadbeat [19], iterative learning [20], repetitive learning method [21], sliding mode [22], and model predictive techniques (MPCs) [23]. Compared to advanced control strategies, the conventional PI control has a simple structure. But due to its sensitivity to parameter uncertainties and external disturbances, it is challenging to suppress the torque ripples effectively [24]. In order to address these issues, PI-based resonant controller was analyzed and compared to traditional PI controllers in [25]. Since intelligent controllers are able to operate under high levels of uncertainty and deal with a huge amount of data, its concept commonly applies to anything that cannot be defined as conventional control. Intelligent control is multidisciplinary since it expands and incorporates theories and techniques from fields including research and operation, computer science, and control [26, 27]. The researchers are working on developing control techniques based on cognitive processes like learning and decision, such as fuzzy systems and NNs [28]. Numerous studies are being conducted to determine how fuzzy logic controllers (FLCs), artificial neural networks (ANNs), and neuro-fuzzy controllers (NFCs) can be applied to PMSM-based drive systems [29]. FL rectifies nonlinearities in the system through the use of human expertise. But, it needs to fine-tune its membership functions (MFs) because it relies too much on the designer's intuition and experience [30]. NFC is rapidly being employed in drive motion control. This controller uses a NN, to account for any parameter variations, and FL, to execute the dynamic behavior [31]. The complex nonlinear properties of the system can be addressed by the NN; however, there is an issue with the lengthy training and convergence durations [32]. By combining fuzzy inference and NNs, an Adaptive Neuro-Fuzzy Inference System (ANFIS) is developed. The ANFIS analyzes how the drive system performs and tracks the target reference by comparing the performance of drive system at different operating points [33, 34]. However, each control method has its own distinct advantages and disadvantages, which are listed in Table 3.1. The PI-RES controller has the capability to introduce infinite gain at the targeted resonant frequencies. The ANFIS offers numerous advantages, such as the ability of rapid learning, adaptability,

Table 3.1 Advantages and disadvantages of different controllers

Types	*Advantages*	*Disadvantages*
Proportional-integral derivative (PID)	Easy to implement & maintain	Mainly used for linear time invariant system
Sliding mode control (SMC)	Used for nonlinear system	Cause error proneness
Predictive control (PC)	Good performance, strong robustness, low requirements for model accuracy	Heavy calculations, not suitable for fast time varying system
Neural network control (NNC)	Convergence to a precise model	Need a longer training time and a slower learning rate to get the accurate model
Fuzzy logic control (FLC)	Simple control structure, good robustness, efficient design, no need for precise models	Lower accuracy and steady-state error

and the capability to capture the nonlinear structure of a system [35, 36]. Therefore, this chapter introduces a novel ANFIS-based resonant (ANFIS-RES) controller. The proposed controller deals with the inherent nonlinearity in PMSM drive and provides quick tracking of the reference signal. The effectiveness of the proposed controller is shown by comparing the traditional proportional-integral resonant (PI-RES) controller response and the ANFIS-RES controller response.

This chapter is divided into five sections, comprising of introduction in Section 3.1. The conventional control strategy is discussed in Section 3.2. Section 3.3 consists of the theoretical explanation of the proposed ANFIS-based control algorithm. The implementation of the proposed technique is described in Section 3.4. Section 3.5 consists of result and discussion. The concluding remarks are described in Section 3.6.

3.2 CONVENTIONAL CONTROL STRATEGY

There are a number of techniques for controlling electrical drives, but the field oriented control (FOC) method for the control of PM motors is popular due to its effectiveness in improving the dynamic performance of drive systems [37].

The equivalent model of PMSM is generated by transforming the motor drive's voltage into its $d-q$ axis parameters. Since the rotor is made up of permanent magnets without any electrical magnetization, the mathematical modeling of a PMSM mostly concentrates on the stator. The necessary $d-q$ stator voltages and flux linkages are represented mathematically by Equations (3.1) to (3.4), respectively.

$$v_d^s = R_s i_d + \frac{d}{dt}\psi_d - \omega_r \psi_q \tag{3.1}$$

$$v_q^s = R_s i_q + \frac{d}{dt}\psi_q - \omega_r \psi_d \tag{3.2}$$

$$\psi_q = L_q i_q \tag{3.3}$$

$$\psi_d = \psi_f + L_d i_d \tag{3.4}$$

where ω_r is the rotor speed; $v_d{}^s$ and $v_q{}^s$ are stator voltages in $d-q$ reference frame; L_d and L_q are inductances on $d-q$ axis; R_s represents the stator resistance; i_d and i_q are current in $d-q$ reference frame; and ψ_d and ψ_q are flux linkages in $d-q$ frame. The electromagnetic motor torque, T_e, produced by the motor is specified by Equation (3.5), where P represents the number of pole pairs.

$$T_e = \left(\frac{3}{2}\right)\left(\frac{P}{2}\right)\left(\psi_d i_q - \psi_q i_d\right) \tag{3.5}$$

The robustness of the control approach and the accuracy in system response play a major role in the efficient performance of PMSM drive systems. The steady-state error is removed using the PI controller but it is very sensitive to changes in speed, parameter fluctuations, and loading circumstances [38].

3.2.1 Traditional PI-RES controller design

The traditional Proportional Resonant (P-RES) controller is generated by combining the proportional and resonant controllers. The s-domain is used to express a P-RES controller for tracking an AC signal, as described in [39]. The transfer function of traditional P-RES controller, $G_{PR}(s)$, is given by Equation (3.6).

$$G_{PR}(s) = K_P + \frac{2K_{ri}\omega_c s}{s^2 + 2\omega_c s + \omega^2} \tag{3.6}$$

where ω is the adjusted signal frequency, ω_c is the cut-off frequency, K_{ri} is the resonant coefficient, and K_P is the proportional constant.

The conventional PI-RES controller is designed by combining PI and resonant controller in a parallel configuration. The combined transfer function of traditional PI-RES, $G_{PI-RES}(s)$, is represented by Equation (3.7).

$$G_{PI-RES}(s) = K_P + K_I + \frac{2K_{ri}\omega_c s}{s^2 + 2\zeta\omega_c s + \omega^2} \tag{3.7}$$

where ξ is the damping coefficient and K_I is the integral controller constant.

A conventional PI controller often offers a satisfactory response for a low-power application, but PI tuning cannot be used to increase drive performance in a high-power application [40]. Considering this drawback in order to achieve higher dynamic performance, the PI controller is replaced by the ANFIS controller in this proposed system.

3.3 PROPOSED CONTROLLER DESIGN

In this section, the resonant controller based on ANFIS is presented. The resonant controller has the advantage of having high gain at the resonant frequency, which makes them capable of reducing steady-state error when tracking or rejecting a sinusoidal input.

Jang et al. introduced the ANFIS approach for the first time in 1993 [41]. The conceptual technique of ANFIS is formed by combining two machine learning schemes of FL and NNs. ANFIS is an elementary data-learning method that employs FL to convert inputs into desired outputs. In a hybrid (Neuro-fuzzy) model, FL and NNs learning algorithms are integrated. The values of parameters are determined by the NN and FL governs the IF-THEN logic [42, 43]. Some of the characteristics, showing the successful implementation of ANFIS in EDS, are listed below:

- It provides the desired data collection, more options of MF and excellent explanation capabilities using fuzzy rules.
- It is simple to apply due to its linguistic and mathematical expertise involved in problem-solving. So, it facilitates quick and precise learning.
- It does not require prior human expertise.

3.3.1 ANFIS architecture

ANFIS is a type of adaptive network similar to Fuzzy Inference System (FIS). In ANFIS, the MF parameters are modified or tuned by using either a hybrid learning algorithm or a back propagation algorithm with the particular input–output data [44, 45]. The output MF cannot be provided by multiple rules. There must be an equal number of rules and MFs. The ANFIS architecture, based on the first-order Sugeno model, is described using two fuzzy IF-THEN rules, as given below, where for inputs x and y, A_1, A_2 & B_1, B_2 are the fuzzy sets for variables x and y whose outputs are denoted by f_1 in which p_1, q_1, r_1 & p_2, q_2, r_2 are the parameters of MF.

- Rule 1: if x is A_1 and y is B_1 then, $f_1 = p_1x + q_1y + r_1$
- Rule 2: if x is A_2 and y is B_2 then, $f_1 = p_2x + q_2y + r_2$

To implement these two rules, the architecture of ANFIS has five layers, as shown in Figure 3.1, where w_1, w_2 are the weights of neurons, $\bar{w}_1, \overline{w_2}$ are normalized weights of neurons. x, y are the inputs, A_1, A_2 are fuzzy sets for variable (x), B_1, B_2 are fuzzy sets for variable y, and f is the output within fuzzy region. These five layers are Fuzzification layer,

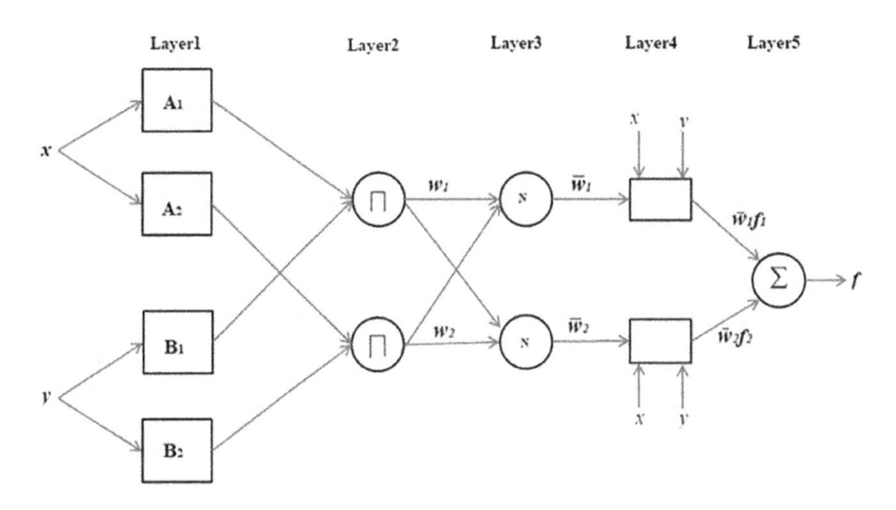

Figure 3.1 Basic structure of ANFIS.

Product layer, Normalization layer, De-fuzzy layer, and Output layer. In this architecture, a square stands for an adaptive node and a circle indicates a fixed node.

A detailed explanation of these layers is mentioned below:

- *Layer 1.* Fuzzification layer
  The node function for layer 1 is given by Equation (3.8). Here, $O_{1,i}$ represents the membership grade of fuzzy set A_1 & A_2. The terms $\mu_{A_i}(x)$ and $\mu_{B_{i-2}}(y)$ represent the fuzzy MF. The subscripts A_i and B_{i-2} represent the relevant linguistic level to this node. The term i denotes the number of nodes.

$$O_{1,i} = \begin{cases} \mu_{Ai}(x) & ; \text{for } i = 1,2 \\ \mu_{Bi-2}(y) & ; \text{for } i = 3,4 \end{cases} \tag{3.8}$$

- *Layer 2.* Product layer
  Each of the nodes in this layer is a circle node with label π, which multiplies the incoming signals and sends the resulting output signal. The node function for layer 2 is given by Equation (3.9). Here, $O_{2,i}$ or w_i represents the firing strength and $\mu_{B_i}(y)$ represents the MF.

$$O_{2,i} = w_i = \mu_{A_i}(x)\mu_{B_i}(y);\ i = 1,2 \tag{3.9}$$

- *Layer 3.* Normalization layer
  An N-labeled circle node characterizes each node in this layer. The i-th node figures out the firing strength of the i-th rule in relation to the total firing strength of all rules. Every node is a fixed node labeled with 'N' and the corresponding node function is given by Equation (3.10). Here, $O_{3,i}$ or $\overline{w_i}$ represents the normalized firing strength of a particular neuron. The terms w_1 and w_2 represent weights of neurons.

$$O_{3,i} = \overline{w_i} = \frac{w_i}{w_1 + w_2}; \; i = 1, 2 \tag{3.10}$$

- *Layer 4.* De-fuzzy layer
  In this layer, node i is always a square node with a node function and every node in this layer is an adaptive node. The corresponding node function is given by Equation (3.11). Here, $O_{4,i}$ represents the normalized strength of output and P_i, q_i, r_i represents the parameter sets of node. The normalized firing strength from layer 3 is represented by w_i .

$$O_{4,i} = \overline{w_i} f_i = \overline{w_i}\left(P_i x + q_i y + r_i\right) \tag{3.11}$$

- *Layer 5.* Output layer
  There is only one node in this layer. It is a circular node, represented by Σ, which calculates the overall output as the sum of all incoming signals. The corresponding node function is represented by Equation (3.12). Here, $0_{5,i}$ represents the overall output and f_i represents output at node i within the fuzzy region.

$$0_{5,i} = \sum \overline{w_i} f_i = \frac{\sum w_i f_i}{\sum w_i} \tag{3.12}$$

The basic ANFIS architecture is now combined with the RES controller technique to form the proposed ANFIS-RES controller. The schematic representation of the proposed ANFIS-RES technique is shown in the form of flow chart in Figure 3.2. The FIS is trained using input–output training data that is taken from the PMSM reference model. 'Two inputs and one output' interface is considered for the design of the ANFIS-RES controller.

For collecting data for training, the error $(e(k))$ between actual motor speed and reference speed and the change in the error $(\Delta e(k))$ are given

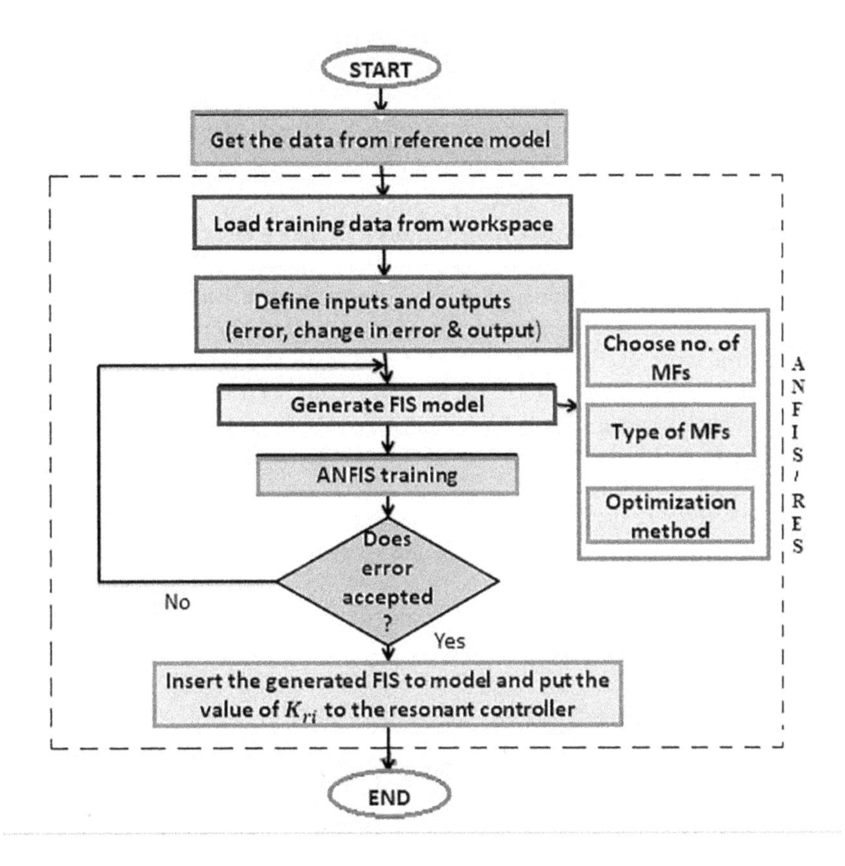

Figure 3.2 Flow chart for the proposed method.

as inputs to the controller. The change in control signal (du) is selected as output. Although there are a lot of errors (inputs) available, only error and change in error are considered as input in the proposed algorithm because the speed response is taken as a reference and as the main aim of FLC is to minimize the error between reference input and actual output. The error and the change in error, in terms of inputs for the proposed controller, are given by Equations (3.13) and (3.14), respectively, where w_{ref} is the reference speed and w_m is the motor speed of PMSM drive.

$$e(k) = w_{\text{ref}} - w_m \tag{3.13}$$

$$\Delta e(k) = e(k) - e(k-1) \tag{3.14}$$

A proper sampling time $(T_s = 10e-6)$ is used to create the collection of training data from the reference model. The data collected from the PMSM reference model is considered for the execution of proposed controller. The next step is determination of input–output and MFs. Additionally, ANFIS-RES controller provides the option of choosing MF types such as triangular, gauss, and trapezoidal [46].

Using the Sugeno model, the number of MFs is selected by recording the error in the settling time of rotor speed. Ideally, seven MFs are selected for implementing ANFIS architecture for any control system [47]. However, the number of MFs is varied to examine the effect on settling time error. Due to this error test, the optimum number of MFs is obtained for which the settling time error is minimum. The error obtained using three, five, and seven MFs are found to be 9.735, 0.0374, and 6.22. It is observed that five triangular MFs are providing the least error. Hence, five MFs are fixed for further implementation of ANFIS-RES controller to the PMSM drive system. There are 5 MFs for error input and 5 MFs for change in error input, so the total rules obtained is 25.

'START' depicts the beginning of control process; 'END' depicts the closure of the control process.

3.3.2 ANFIS-RES algorithm

The different steps that are considered for the design of controller are as follows:

Step 1. Collect the data from different inputs, change in error $(\Delta e(k))$ and error $(e(k))$, the output (du), and provide it as loading data to train ANFIS. The error at a specified number of epochs is shown in Figure 3.3 and it is observed that the error settles down after epoch 35 and is near to 0.0374. So, this value

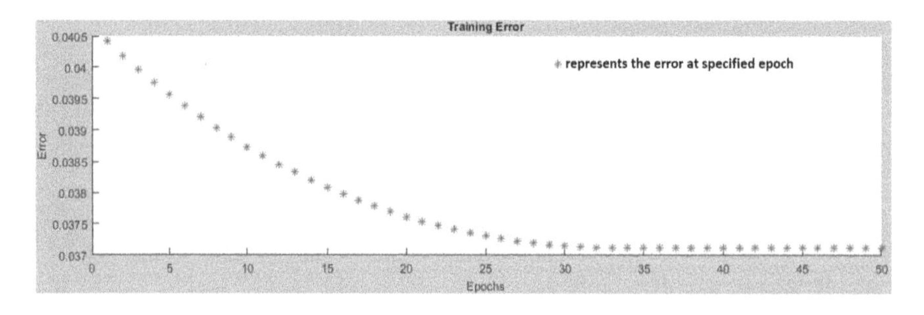

Figure 3.3 ANFIS training data.

of error is considered for the implementation of the proposed technique.

Step 2. The five MFs for each of the two inputs that have been selected are illustrated in Figure 3.4. It shows that triangular MFs vary in the range of 0–1.

Step 3. The proposed controller regulates the motor speed while taking the rules into consideration. Figure 3.5 represents the ANFIS-RES layers created using MATLAB® Simulink tool, employing five layers.

Step 4. Using grid partition, the FIS is generated, and then the hybrid method is selected for training the loaded data.

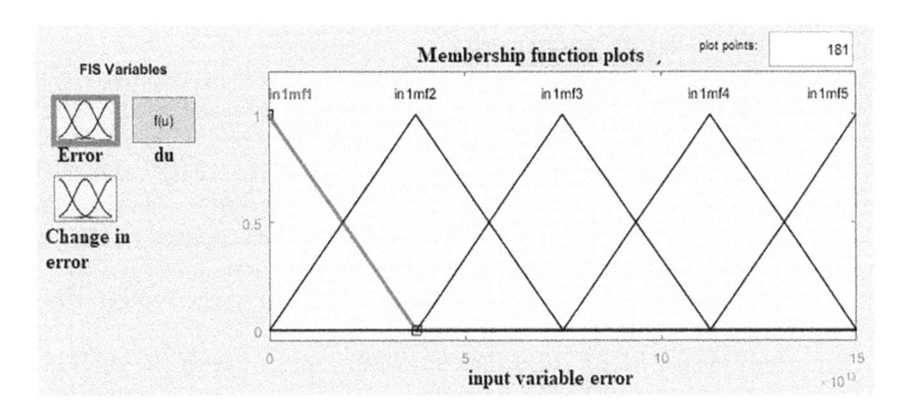

Figure 3.4 Membership functions.

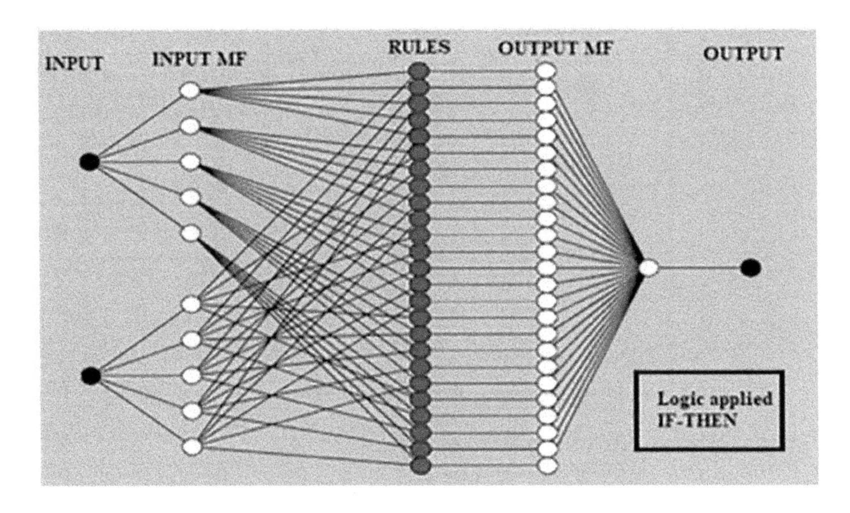

Figure 3.5 Five-layered ANFIS architecture.

Step 5. After completion of ANFIS training, the obtained result at output layer is added to the Resonant controller output that also requires computation of resonant coefficient, K_{ri}. Thus, the combined output of ANFIS-RES controller is utilized by the drive system in order to generate the necessary inverter pulses.

3.4 IMPLEMENTATION OF ANFIS-RES CONTROLLER

Here, in this section, first, the implementation of FOC-based control strategy is explained for a PMSM drive system. Thereafter, the proposed ANFIS-RES controller is implemented. The corresponding designed MATLAB® Simulink models are also presented.

A typical closed loop vector control strategy for PMSM-based drive systems is illustrated in Figure 3.6. The torque limiter uses measured and referenced speeds to calculate the reference value of the current i_q. The reference values of currents, i^*_{abc}, is calculated by applying a reverse Park's transformation using the rotor position, i_d and i_q. The performance criterion that is utilized in this research to assess the feasibility of the proposed technique for minimizing torque ripples is the Torque Ripple Factor (TRF). TRF is defined as the peak to peak torque to the rated torque of the PMSM drive system.

The traditional PI-RES controller has been replaced with the ANFIS-based resonant controller. The proposed ANFIS-RES controller block is

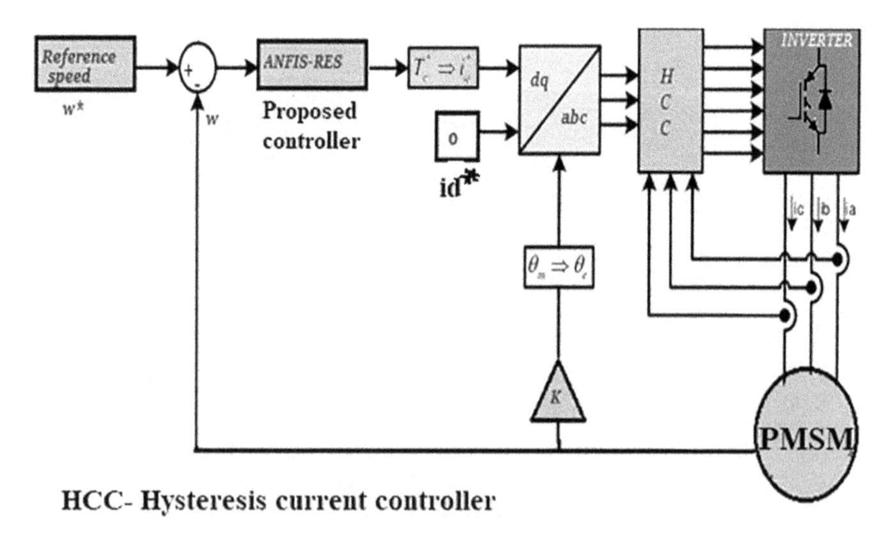

Figure 3.6 Schematic diagram of the PMSM-based drive system, using the proposed ANFIS-RES-based control scheme.

presented in an expanded form in Figure 3.7, where K_{ri} is the resonant coefficient term and T_s is the sampling time. The different parameters for training the proposed technique are shown in Figure 3.8. The ANFIS rules generated are shown in Figure 3.9.

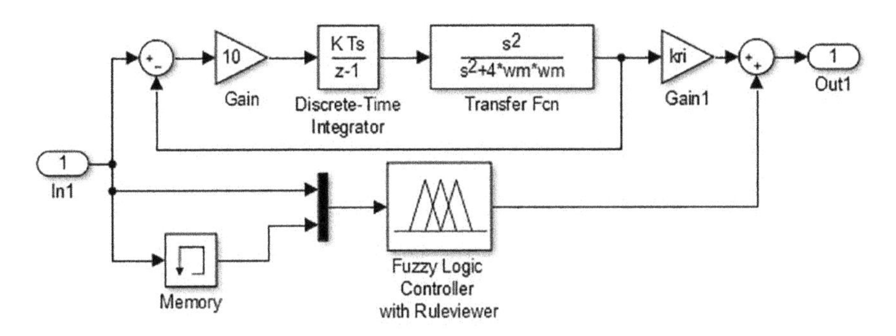

Figure 3.7 Expanded view of the proposed ANFIS-RES controller block.

```
ANFIS info:
    Number of nodes: 75
    Number of linear parameters: 25
    Number of nonlinear parameters: 30
    Total number of parameters: 55
    Number of training data pairs: 2001
    Number of checking data pairs: 0
    Number of fuzzy rules: 25
```

Figure 3.8 Parameters involved in the training of the proposed controller.

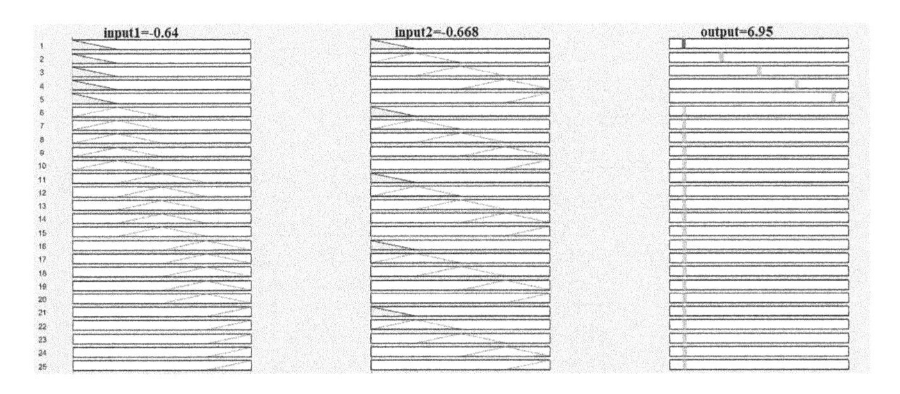

Figure 3.9 ANFIS rule viewer where 25 rules are depicted where 'input1' is the error in speed; 'input2' is the change in error of the speed response, and 'output' is the q-axis current.

A simulation model of the PM motor drive has been developed using MATLAB®/Simulink based on the parameters of the simulated PMSM in Table 3.2. The Simulink model of the proposed ANFIS-RES control scheme is shown in Figure 3.10.

3.5 RESULTS AND DISCUSSION

This section presents the results and corresponding remarks for the proposed controller design. First, the PI-RES-based controller output is presented for PMSM-based drive systems. Then, the output response of the proposed ANFIS-RES controller is reported for the same system configuration. For both controller designs, the transient responses are recorded and presented at no load and rated load conditions respectively.

3.5.1 Performance analysis using the PI-RES controller

The PI-RES controller generates the reference current based on the error in the reference speed and motor speed in the speed loop. This controller

Table 3.2 PM synchronous motor parameters

Parameters	*Ratings*	*Units*	*Parameters*	*Ratings*	*Units*
Rated speed	1500	rpm	No. of poles	8	-
Rated torque	11	Nm	Flux	0.865	Wb
Rated voltage	415	V	d-axis inductance	0.0114	H
Rated current	6.9	A	q-axis inductance	0.0114	H
Stator resistance	1.93	Ω	Inertia	0.11	$kg\ m_2$

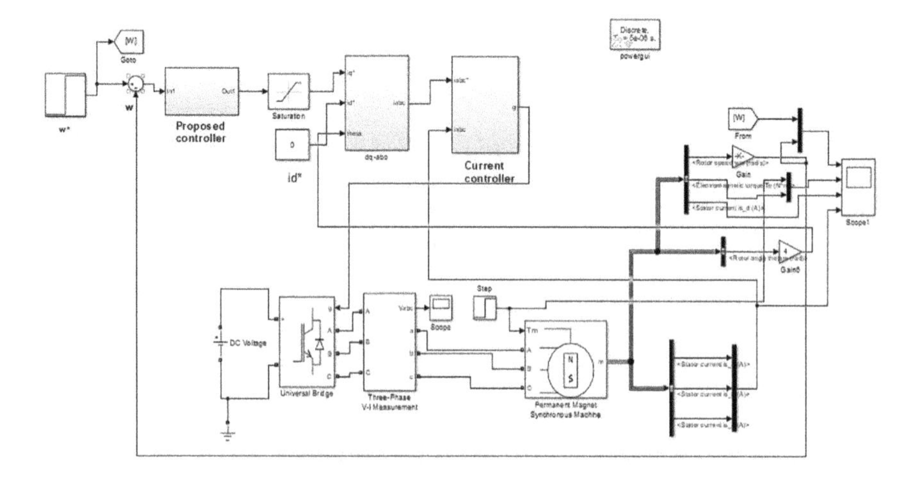

Figure 3.10 Simulink model of PMSM using the proposed control technique.

requires the tuning of parameters K_p, K_I & w_c. Using the PI-RES controller, the responses of the PMSM are analyzed at 1500 rpm and under varying load conditions. In Figure 3.11, the performances of rotor speed, torque, and stator current at no load are presented. The full load analysis results are presented in Figure 3.12. The motor settles to the rated speed of 1500 rpm in 0.012 sec and tracks the reference speed smoothly when there is no applied load.

When rated load (11 Nm) is introduced at 0.03 sec, the speed response drops to 1465 rpm (a drop of 35 rpm) and settles to the rated speed at 0.0331 sec, but after that the drive then smoothly tracks the reference speed.

3.5.2 Performance analysis using the ANFIS-RES controller

The ANFIS-RES controller uses the collected speed data from reference model and K_{ri} value to analyze the performance of PMSM-based drive. The value of K_{ri} is set to 25.123. The simulation result of PMSM-based drive using the proposed controller is illustrated in Figure 3.13 at no load condition and in Figure 3.14 at rated load condition. It is observed that while running the motor at no load, the time taken by the rotor speed to settle up to 1500 rpm is 0.01 sec, which is slightly smaller (~0.2%) than that

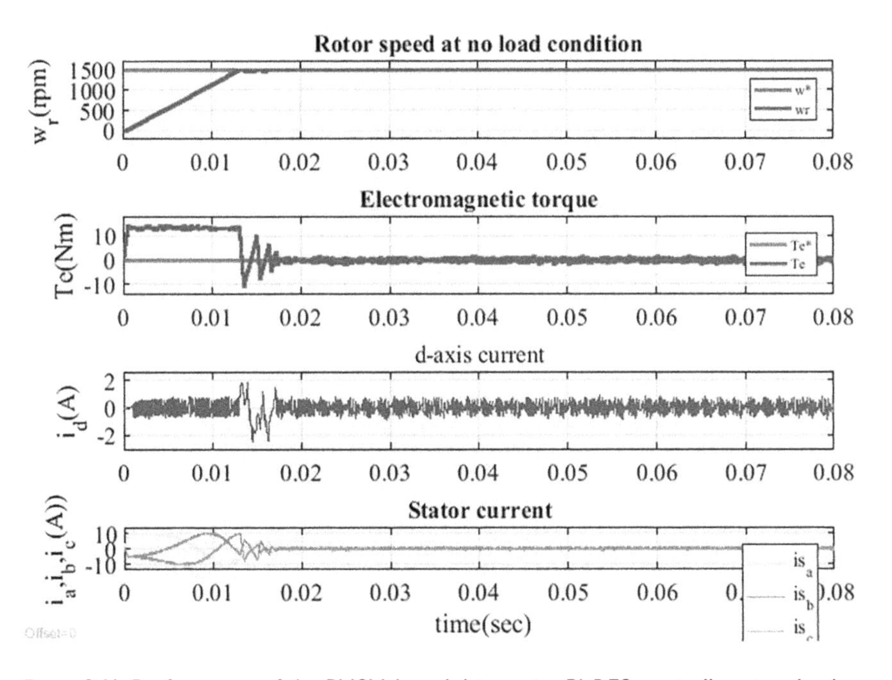

Figure 3.11 Performance of the PMSM-based drive using PI-RES controller at no load.

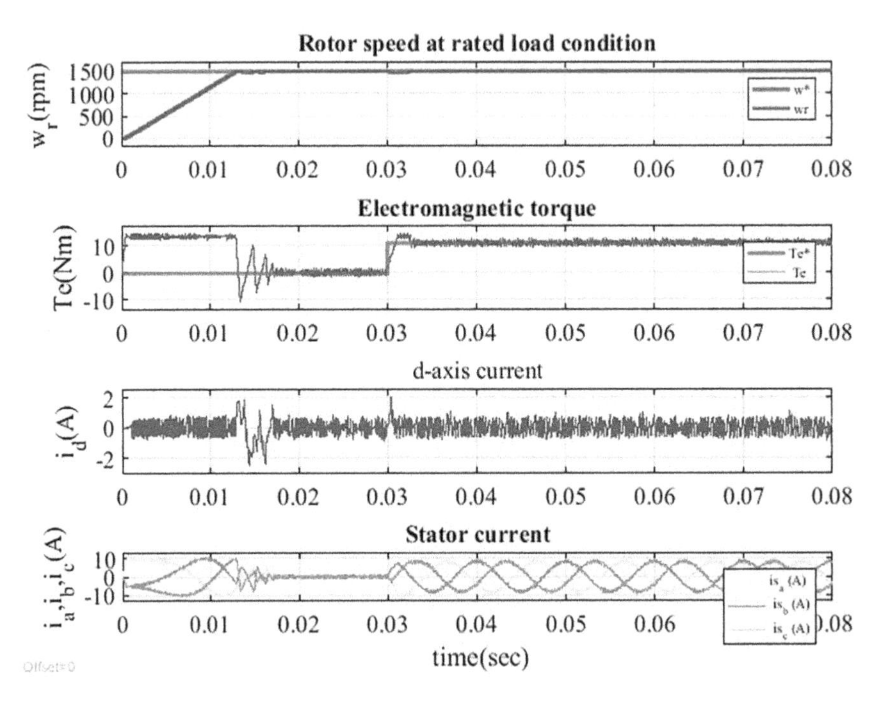

Figure 3.12 Performance of the PMSM-based drive using PI-RES controller at rated load condition.

obtained by the PI-RES controller. Similarly, the torque and phase currents are quickly settling to steady-state value instantly.

Thereafter, the ANFIS-RES-based PMSM drive system is run at rated load condition. Initially, at time (t=0 sec), the motor is run at zero load condition. Then at 0.03 sec, a load disturbance of 11 Nm is introduced in the drive system. At this instant, it is observed that, first, the rotor speed drops up to 1487 rpm (a drop of 13 rpm) and then settles to the rated speed of 1500 rpm in 0.0311 sec. Also, it is observed that using the proposed controller, the ripples in torque are reduced by 4% and the settling time of the rotor speed is reduced by ~6%, as compared to that obtained by the PI-RES controller.

3.5.3 Comparative analysis of the proposed controller and the PI-RES controller

This section presents the comparison of both control schemes. The following observations are recorded while performing the comparative analysis of the PMSM drive system, when it is tested using both control schemes separately:

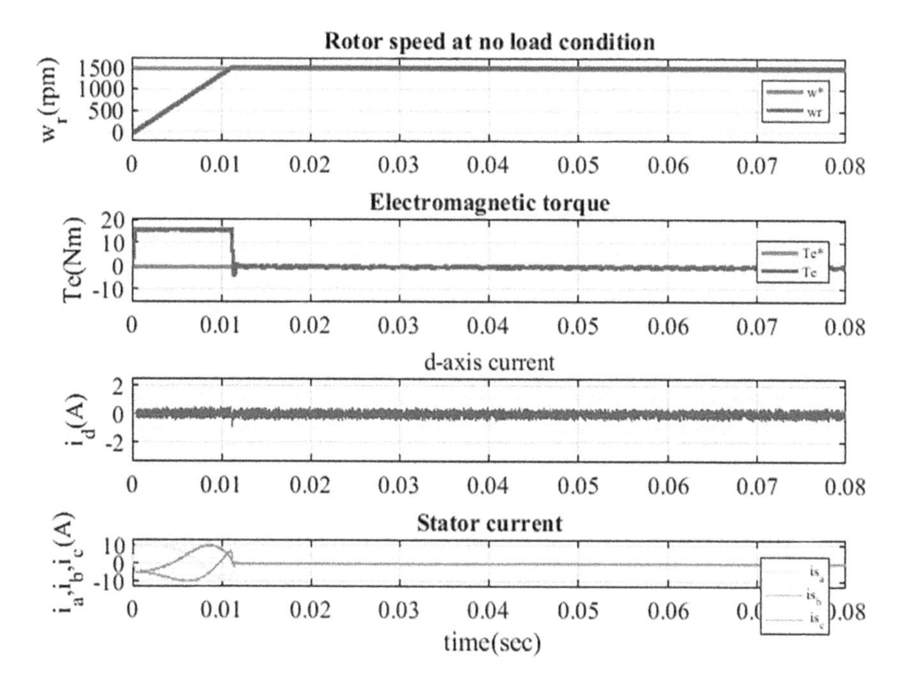

Figure 3.13 Performance of the proposed controller at no load.

- Using the PI-RES controller scheme, the settling duration of rotor speed and ripples in torque is larger than that obtained using the proposed controller.
- In the traditional PI-RES controller, parameters like K_p, K_I & w_c are needed to be adjusted. This increases the sluggishness in the speed and torque response.
- While in the ANFIS-RES controller, the requirement of control parameters adjustment is eliminated.

The comparison of the speed and torque performance of PMSMs is shown in Figure 3.15 employing the PI-RES controller and in Figure 3.16 using the ANFIS-RES controller. The analysis shows that the TRF is reduced by 4% using the proposed controller. When load torque is applied at the instant of 0.03 sec, the speed drops by 13 rpm using the ANFIS-RES controller and by 35 rpm using the PI-RES controller, respectively. Then, to again attain the rated speed, the time taken by speed response is 0.0331 sec (using PI-RES) and 0.0311 sec (using the proposed controller).

The overall comparison between two control strategies is presented in Table 3.3. Thus, it is observed that, at no load as well as rated load

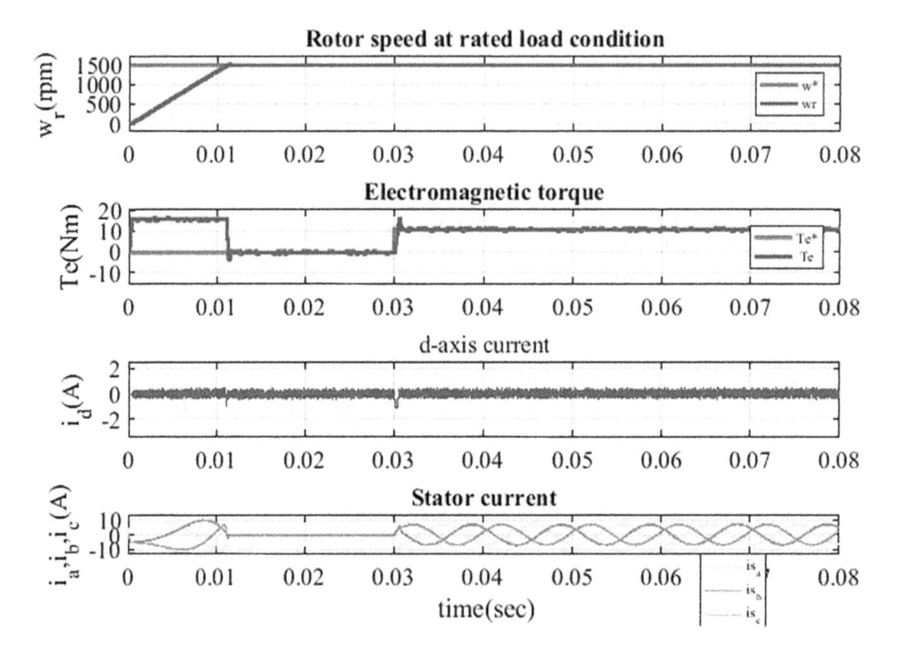

Figure 3.14 Performance of the proposed controller at rated load condition.

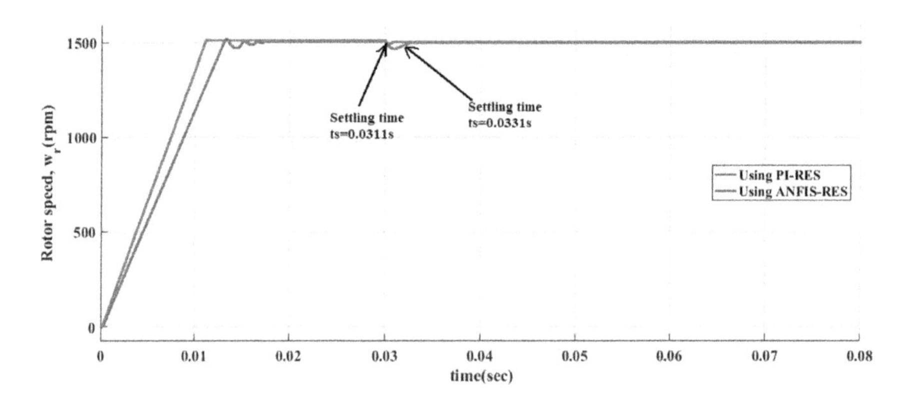

Figure 3.15 Comparison of speed response using both the techniques.

condition, using ANFIS-RES controller, PMSM drive system gives a better performance.

Generally, MTS pay more attention toward the resolution of the issues caused due to traffic congestion, safety concerns, and pollution [48]. The implementation of ICS technologies is expected to add its relevant contribution, in improving the performance of MTS. It resolves such issues by incorporating rapid decision-making capability in its control system. Modern

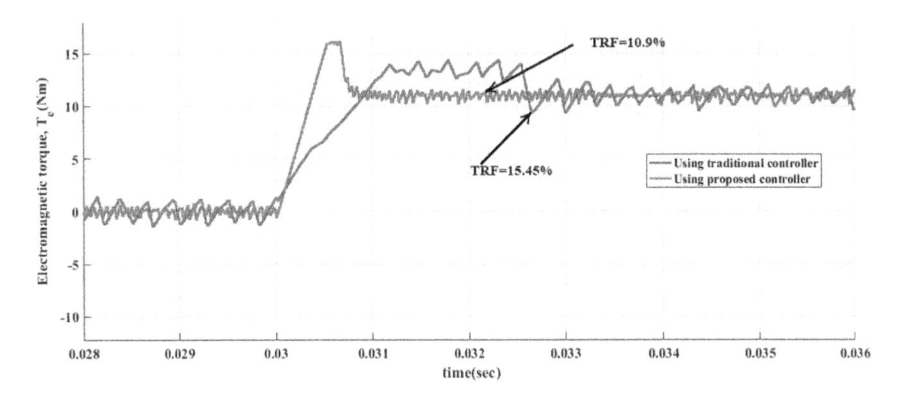

Figure 3.16 Comparison of torque response using both the techniques.

Table 3.3 Comparative analysis of ANFIS-RES and PI-RES control scheme

Scheme	*Speed settling time (sec) (at no load condition)*	*Drop in rotor speed at the instant of load disturbance (rpm)*	*Time taken to settle to 1500 rpm after the load disturbance (sec)*	*Motor torque ripple factor, TRF (%)*
PI-RES	0.012	35	0.0331	15.45
ANFIS-RES	0.010	13	0.0311	10.90

transportation systems (MTS) with ICS technologies are able to monitor, assess, and manage them without getting affected by any fluctuation in system parameters [49]. As it is shown by this study, the dynamic performance of PMSM-based drive systems has been enhanced. PMSM-based drives are one of the modern powertrain systems in the transportation industry. Hence, the coalition of ICS schemes has appreciable advantages in MTS technologies.

3.6 CONCLUSION

In this chapter, a traditional controller and an ANFIS-based controller are designed and analyzed. Thereafter, both control strategies are compared to check the performance of PMSM-based drive systems. Using the proposed ANFIS-RES controller, the ripples in torque, introduced due to nonlinearities present in the system, are reduced which increases the performance of PMSM drive systems. For the implementation of the proposed controller, the training data from a PMSM dynamic model is taken. The proposed controller has the advantage of not being affected by any change in system parameters such as load disturbances. To assess its dynamic performance, the FOC of PMSM-based drive is analyzed under no load and rated load circumstances because these two conditions are the worst scenarios where the PMSM drive system needs to perform well. It is seen that the ripples in torque response of PMSM under transient condition are minimized using the ANFIS-RES controller by adjusting the resonant coefficient term. The outcomes of performed studies illustrate that the proposed controller gives better dynamic performance. For example, the TRF of the system is reduced by 4%, speed drop is reduced by 62%, and the time of speed settlement is decreased by 6.04% during transient condition. However, training the data in the proposed controller takes more time. The selection of the number and type of MFs has an impact on the training error. These limitations may be utilized for future studies for the extension of the proposed work.

REFERENCES

[1] Vas, Peter, *Artificial Intelligence-Based Electrical Machines, and Drives*, Oxford University Press, 1999.
[2] A. Zilouchian, and M Jamshidi, *Intelligent Control Systems Using Soft Computing Methodologies*, CRC. 2011
[3] S. Rajasekaran, and G.A. Vijayalakshmi Pai, *Neural Networks, Fuzzy Logic, and Genetic Algorithms: Synthesis and Applications*, Published with *"PHI Learning Private Limited"* in 2013.
[4] M. N. U. Md, and M. Islam, "Development and implementation of a new adaptive intelligent speed controller for IPMSM drive", *IEEE Trans. Ind. Appl.*, vol. 45, no. 3, pp. 1844–1851, May/June 2009.
[5] P. Santhosh, R.H. Chile, and A.B. Patil, "Model reference adaptive technique for sensorless speed control of induction motor", *First International Conference on Emerging Trends in Engineering and Technology*, pp. 893–898, July 2008.
[6] T.S. Gabbi, H.A. Gründling, and R.P. Vieira, "Sliding mode MRAS speed observer applied to permanent magnet synchronous motor with decoupled current control", *IECON 2016–42nd Annual Conference of the IEEE Industrial Electronics Society*, pp. 2929–2934, 2016 October.
[7] S.V. Emel'yanov, "Theory of variable-structure control systems: Inception and initial development", *Comput. Math. Model.*, vol. 18, pp. 321–331, 2007. https://doi.org/10.1007/s10598-007-0028-6
[8] H. Kim, J. Hartwig, and R. D. Lorenz, "Using on-line parameter estimation to improve efficiency of IPM machine drives,' ' in *Proc. IEEE 33rd Annu. IEEE Power Electron. Spec. Conf.*, pp. 815–820, Jun. 2002.
[9] J.S. Park, B.-G. Gu, J.-H. Kim, J.-H. Choi, and I.-S. Jung, "Fabrication of Sensorless Drive for Automotive Applications, " *4th International Conference on Power Engineering, Energy and Electrical Drives*, pp. 187–191, 2013, doi: 10.1109/PowerEng.2013.6635604.
[10] P. Pillay, and R. Krishnan, "Modeling, simulation and analysis of permanent magnet motor drives- Part II: The Brushless DC motor drive", *IEEE Trans., Ind. Appl.*, vol. 25, no. 2, pp. 274–279, 1989

[11] A. Senthil Kumar, T. Prasath Vijay Raj, A. Tharagesh, and V. Prasanna, "Design and analysis of a permanent magnet DC motor," *Lect. Notes Electr.* Eng., vol. 442, pp. 237–249, 2018, doi: 10.1007/978-981-10-4762-6_22.
[12] M. Tuna, C. B. Fidan, S. Kocabey, and S. Görgülü, "Effective and reliable speed control of permanent magnet DC (PMDC) motor under variable loads," *J. Electr. Eng. Technol.*, vol. 10, no. 5, pp. 2170–2178, 2015, doi: 10.5370/JEET.2015.10.5.2170
[13] R. Islam, I. Husain, A. Fardoun, and K. McLaughlin, "Permanent-magnet synchronous 168 motor magnet designs with skewing for torque ripple and cogging torque reduction," *IEEE Trans. Ind. Appl.*, vol. 45, no. 1, pp. 152–160, 2009, doi: 10.1109/TIA.2008.2009653.
[14] N. P. Quang, and J.-A. Dittrich, *Vector Control of Three-Phase AC Machines*, 1st ed. *Springer-Verlag Berlin Heidelberg*, 2008.
[15] D. Xu, B. Wang, G. Zhang, G. Wang, and Y. Yu, "A review of sensorless control methods for AC motor drives," *CES Trans. Electr. Mach. Syst.*, vol. 2, no. 1, pp. 104–115, 2020.
[16] A. Gebregergis, M. H. Chowdhury, M. S. Islam, and T. Sebastian, "Modeling of permanent-magnet synchronous machine including torque ripple effects," *IEEE Trans. Ind. Appl.*, vol. 51, no. 1, pp. 232–239, 2015, doi: 10.1109/TIA.2014.2334733.
[17] Y. Zhang, W. Cao, S. McLoone, and J. Morrow, "Design and flux-weakening control of an interior permanent magnet synchronous motor for electric vehicles", *IEEE Trans. Appl. Supercond.*, vol. 26, no. 7, pp. 1–6, 2016, Art no. 0606906, doi: 10.1109/TASC.2016.2594863.
[18] Ch. M. Rao, B. M. Krishna, A.L. Soundarya, and N.K. Kumari, "FOC of PMSM with Model Reference Adaptive Control Using Fuzzy-PI Controller", *IJCTA*, vol. 8, No. 1, Jan-June 2015, pp. 96–108.
[19] H. Tang, and W. Zhao, "Deadbeat control of permanent magnet synchronous motor based on sliding active disturbance rejection controller", *2015 18th International Conference on Electrical Machines and Systems (ICEMS), 2015*, pp. 1402–1406, doi: 10.1109/ICEMS.2015.7385258.
[20] Weizhe Qian, S.K. Panda, and Jian-Xin xu, "Torque ripple minimisation in PM synchronous motor using iterative learning control", *IEEE Trans. Power Electron.*, vol. 19, no. 2, pp. 272–279, March 2004.
[21] K. Jin, M. Sun and Y. Ye, "Robust repetitive learning control for a class of time-varying nonlinear systems", *Proc. 29th Chin. Control Conf.*, vol. 11, pp. 2071–2076, 2010.
[22] S. Chi, and L. Xu, "Position sensorless control of PMSM based on a novel sliding mode observer over wide speed range," in *Proc. IEEE International Power Electronics and Motion Control Conf.*, Shanghai, China, Aug. 2006, vol. 3, pp. 1–7.
[23] H. Zhu, X. Xiao, and Y. Li, "Torque ripple reduction of the torque predictive control scheme for permanent-magnet synchronous motors", *IEEE Trans. Ind. Electron.*, vol. 59, no. 2, pp. 871–877, Feb. 2012.
[24] F. Colamartino, C. Marchand, and A. Razek, "Torque ripple minimization in permanent magnet synchronous servo drive", *IEEE Trans. Energy Convers.*, vol. 14, no. 3, pp. 616–621, 1999.

[25] J. Gao, X. Wu, S. Huang, W. Zhang, and L. Xiao, "Torque ripple minimization of permanent magnet synchronous motor using a new proportional resonant controller", IET Power Electr., vol. 10, no. 2, pp. 208–214, 2017.

[26] M. N. U. Md, and M. Islam, "Development and implementation of a new adaptive intelligent speed controller for IPMSM drive", *IEEE Trans. Ind. Appl.*, vol. 45, no. 3, May/June 2009.

[27] A. Kusagur, S. F. Kodad, and B V. Sankar Ram, "Modeling, design & simulation of an Adaptive Neuro- Fuzzy Inference System (ANFIS) for speed control of induction motor", *Int. J. Comp. Appl.*, vol. 6, no. 12, pp. 1143–1150, September 2010.

[28] C. Xia, C. Guo, and T. Shi, "A neural-network-identifier and fuzzy-controller-based algorithm for dynamic decoupling control of permanent-magnet spherical motor," *IEEE Trans. Ind. Electron.*, vol. 57, no. 8, pp. 2868–2878, Aug. 2010.

[29] M. T. Hagan, H. B. Demuth, and M. H. Beale, "*Neural network design*", Boston, MA: PWS, 2002.

[30] H. -Y. Chung, C. -C. Hou, and C. -L. Chao, "Speed-control of a PMSM based on Integral-Fuzzy control," 2013 *International Conference on Fuzzy Theory and Its Applications (IFUZZY)*, 2013, pp. 77–82, doi: 10.1109/iFuzzy.2013.6825413.

[31] W. Tong-xu, and M. Hong-yan, "The research of PMSM RBF neural network PID parameters self-tuning in elevator", *The 27th Chinese Control and Decision Conference (2015 CCDC)*, 2015, pp. 3350–3354.

[32] Limei Wang, and Mingxiu Tian, "Study of Fuzzy Controller Based on Neural-network for PMSM Speed Adjustment System, " *2006 6th World Congress on Intelligent Control and Automation*, pp. 3762–3766.

[33] W. A. A. Salem, G. F. Osman, , & S. H. Arfa, "Adaptive Neuro-Fuzzy Inference System Based Field Oriented Control of PMSM & Speed Estimation", in *Twentieth International Middle East Power Systems Conference*, Cairo, 2018, pp. 626–631.

[34] D. Flieller, N.K. Nguyen, P. Wira, G. Sturtzer, D.O. Abdeslam, and J. Merckle, "A self-learning solution for torque ripple reduction for nonsinusoidal permanent-magnet motor drives based on artificial neural networks," *IEEE Trans. Ind. Electron.*, vol. 61, no. 2, pp. 655–666, Feb. 2014.

[35] Hidayat, S. Pramonohadi, Sarjiya, and Suharyanto, "A comparative study of PID, ANFIS and hybrid PID ANFIS controllers for speed control of Brushless DC Motor drive", *2013 International Conference on Computer, Control, Informatics and Its Applications (IC3INA)*, 2013, pp. 117–122.

[36] C. Xia, B. Ji, and Y. Yan, "Smooth speed control for low-speed high-torque permanent-magnet synchronous motor using proportional– integral–resonant controller," *IEEE Trans. Ind. Electron.*, vol. 62, no. 4, pp. 2123–2134, Apr. 2015.

[37] V. Biyani, J. R, T. E. T. A, S. S. V. S, and P. K. P, "Comparative Study of Different Control Strategies in Permanent Magnet Synchronous Motor Drives, " *2021 IEEE 5th International Conference on Condition Assessment Techniques in Electrical Systems (CATCON)*, 2021, pp. 275–281.

[38] Shin-Hung Chang and Pin-Yung Chen, "Self-tuning gains of PI controllers for current control in a PMSM", *2010 5th IEEE Conference on Industrial Electronics and Applications*, 2010, pp. 1282–1286.

[39] A.G. Yepes, F.D. Freijedo, O. Lopez, et al. "High performance digital resonant controllers implemented with two integrators", *IEEE Trans. Power Electron.*, vol. 26, no. 2, pp. 563–576, 2011.
[40] D. Yadav, , and A. Verma, "Behaviour Analysis of PMSM Drive using ANFIS Based PID Speed Controller", in *5th IEEE Uttar Pradesh Section International Conference on Electrical, Electronics and Computer Engineering*, Uttar, 2018.
[41] J.-S. R. Jang, "ANFIS: Adaptive-network-based fuzzy inference system," IEEE Trans. Systems, *Man Cybernetics*, vol. 23, no. 3, pp. 665–684, May 1993.
[42] A. Al-Hmouz, Jun Shen, R. Al-Hmouz, and Jun Yan, "Modeling and simulation of an adaptive Neuro-Fuzzy Inference System (ANFIS) for mobile learning", *IEEE Trans. Learn. Technol.*, vol. 5, no. 03, pp. 226–237, 2012.
[43] Alejandro A. Torres-García, Carlos A. Reyes-García, Luis Villaseñor-Pineda, and Omar Mendoza-Montoya, "Bio signal Processing and Classification Using Computational Learning and Intelligence", *Academic Press*, 2022, pp. 153–176, ISBN 9780128201251.
[44] Jyh-Shing Roger Jang, Chuen-Tsai Sun, and Eiji Mizutani, "*Neuro-Fuzzy and Soft Computing: A Computational Approach to Learning and Machine Intelligence*", MATLAB Curriculum Series. Upper Saddle River, NJ: Prentice Hall, 1997.
[45] B. B. Jovanovic, I. S. Reljin, and B. D. Reljin, "Modified ANFIS architecture - improving efficiency of ANFIS technique", *7th Seminar on Neural Network Applications in Electrical Engineering, 2004. NEUREL 2004*, pp. 215–220, doi: 10.1109/NEUREL.2004.1416577.
[46] R. Ushanandhini, "Adaptive neuro fuzzy inference system with self turning for permanent magnet synchronous motor", *Int. J. Emerg. Technol. Eng. Res. (IJETER)*, vol. 4, no. 3, March 2016.
[47] T. S. Radwan, and M. M. Gouda, "Intelligent speed control of permanent magnet synchronous motor drive based-on neuro-fuzzy approach", *2005 International Conference on Power Electronics and Drives Systems*, 2005, pp. 602–606, doi: 10.1109/PEDS.2005.1619757.
[48] C. H. Fleming, and N. G. Leveson, "Early Concept Development and Safety Analysis of Future Transportation Systems, " in *IEEE Transactions on Intelligent Transportation Systems*, vol. 17, no. 12, pp. 3512–3523, Dec. 2016, doi: 10.1109/TITS.2016.2561409.
[49] S. Ravi, and M. R. Mamdikar, "A Review on ITS (Intelligent Transportation Systems) Technology," *2022 International Conference on Applied Artificial Intelligence and Computing (ICAAIC)*, 2022, pp. 155–159, doi: 10.1109/ICAAIC53929.2022.9792638.

Chapter 4

IoT-based vehicle tracking and accident alert system

Dinesh Kumar Singh, Niraj Kumar Shukla, and Pavan Kumar Singh
Shambhunath Institute of Engineering and Technology

CONTENTS

4.1 INTRODUCTION

In India, every year there is an increase in the number of automobiles on the road and for a busy city on highways; people are rushing to their workplaces using transportation system. It is very difficult to solve these problems related to vehicle accidents and its hazardous consequences. There is an increase in the death rate as well as disabilities due to such accidents. The World Health Organization has released the data that every year nearly 1.35 million causalities cases have occurred across the whole world [1]. These accidents are due to numerous reasons such as fast driving of vehicles, driver's fatigue or the presence of stray animals on the road.

Nobody can be aware of when and where an accident may occur but the life of accident victim impact on the response of the nearby people. Now the main problem is that it is the delayed response of the emergency or medical services which made these accidents fatal. Hence, efforts are being made to decrease the casualties by providing medical assistance to the people within time with the help of modern information and communication technology systems. One such tool of the information system is Internet of Things (IoT), which is used to provide the accident information to the emergency services within time to save human lives [1,2].

The IoT is a concept in which each thing is assigned an internet protocol (IP) address that may be used to find the device on the Internet. Unique identifiers (UIDs) are assigned to mechanical and digital devices so that

DOI: 10.1201/9781003436089-4

data may be sent over the network without the requirement for a human-to-human or human-to-computer connection [3]. LoRaWAN (long-range wide area network), gives the process information for smart application-based on IoT architecture, that is used for this system. This has enough capacity and communication range with low power consumption and cost [3]. In fact, it was initially dubbed as the "Internet of Computers." According to studies, "things" or devices will proliferate at an exponential pace in the next years. Bluetooth and Wi-Fi, two of the most recent advancements in wireless control technology, have enabled the connection of many sorts of devices. Using the Wi-Fi Shield as a mini web server for the Arduino removes the need for physical connections between the board and the computer (or device to device), lowering expenses and enabling the Arduino to operate as an independent device [4,5]. The use of IoT enables to send information automatically regarding the accident to the selected members sufficiently within time. This is due to the reason that the passengers of the vehicle may be unable to call or send a message [6,7].

In this proposed model, the occurrence as well as the exact location of the accident has been detected successfully, with the help of sensors, Global Positioning System (GPS) and Global System for Mobile Communication (GSM) modules [8]. The GPS module is used for detecting the location of the vehicle, and the GSM module is used in the model for sending the messages to the selected members. In case of any false detection of accident or any malfunctioning of the system, a small duration of time has been provided, during which the passenger can cancel the call so that there is no panic situation among the selected members. In [9], GPS and GSM modules are used for the vehicle tracking system. The authors of [10] incorporated the applications of IoT, which made the communication devices to track the vehicle. The model includes GPS and GSM along with a prediction algorithm.

The proposed arrangement is on the basis of a new modern technology. The main motive of this chapter is to intimate about accident as well as its location and send an alert message to the selected members so that the victim can be provided help or support for the survival. This proposed methodology performs both vehicle tracking and accident alert systems, which proves its validity and usefulness.

4.2 INTERNET OF THINGS

The term "Internet of Things" is a combination of different technologies, machine learning, embedded systems and commodity sensors. Data can be transmitted over the network using IoT and the devices can be controlled, which is a system of devices connected to the network provided by UIDs. This made it easier to operate the tool without actually having to deal with it. Networks, sensors, big data and artificial intelligence are used in the state-of-the-art automation and analytics system in IoT. When applied to any sector or system, these solutions provide greater transparency, control and performance [3]. The most usable application ground of IoT are Bluetooth and Wi-Fi.

The following are the features of IoT:

A. ***Intelligence:*** It comes with a combination of IoT algorithms, processing, software and hardware. This improves ambient intelligence capabilities in IoT, which enables creatures to interact intelligently with a given scenario and to perform specific functions. Despite the popularity of smart technologies, IoT intelligence can only be considered a tool for device-to-device interaction, whereas the user–device interaction is managed through simple input methods and a graphical user interface (GUI).
B. ***Connectivity:*** The IoT is made possible by connecting common devices. For the IoT to function as collective intelligence, devices must be able to communicate with each other. Objects can be accessed and used on the network. Networking of smart devices and applications is made possible by this connectivity, which opens up new markets for the IoT.
C. ***Enormous Scale:*** A lot more gadgets presently interface and speak with one another than are accessible on the Internet. The information the board and understanding given by these gadgets is turning out to be more significant.
D. ***Heterogeneity:*** The IoT is one of its most important features: diversity. As a result, devices on the IoT can communicate with each other over multiple networks. The IoT architecture must be able to connect diverse networks directly. Interoperability, modularity and scalability are the four pillars of IoT architecture for diversity systems.
E. ***Security:*** Devices connected to the IoT are vulnerable to security breaches internally. The IoT has many benefits, except the security risks associated with it. The IoT raises many questions about privacy and confidentiality.

4.3 METHODOLOGY

Several studies have been done to detect vehicle accident, but somewhere they were not accurate as they depend on the external data input such as traffic conjunction or by image processing which captures the image of the damaged vehicle. All these needs external input for data transfer. In the proposed system, a novel method is used which will detect the occurrence of accident automatically and send the information of accident to the members. The complete setup contains an accelerometer sensor, and GPS and GSM modules. All these will work collectively to detect accidents and send message about location of the accident to the members. The main objective of the chapter is to save human life by notifying the members that an accident has occurred at a particular location [11]. This results in providing medical help to the victim as soon as possible.

The block diagram of the proposed system consists of electrical and electronic components. These components are integrated as per block diagram and depicted in Figure 4.1. When the occurrence of any accident is detected by an accelerometer sensor, the GPS will note the current position or coordinates of the vehicle. The location of these coordinates is processed further to the microcontroller, which passes this information to the GSM unit. With the help of the GSM unit, we can send the message to the members or medical service providers. The data or information transfer takes place with the

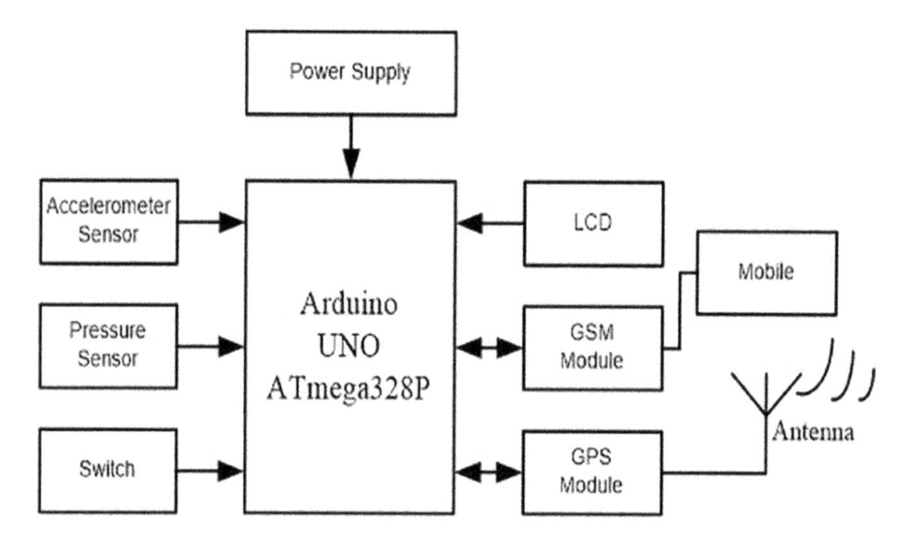

Figure 4.1 Block diagram of system [19].

Table 4.1 Components used in the model

S. No.	*Item*	*Range*	*Model/Size*	*Quantity*
1.	Gear motor	0–12V DC	–	4
2.	Arduino UNO	–	ATmega328P	1
3.	GSM	35 km	SIM800L	1
4.	GPS	3 m position accuracy (0.25 Hz–1 kHz)	GY-6MV2	1
5.	Accelerometer	±3 g (gyro) full scale	ADXL335	1
6.	Battery	12V	–	2
7.	Wheel	–	Diameter-7 cm	4
8.	Connecting wires	–	–	20

help of serial communication between microcontroller, and GSM and GPS modules. In order to receive the message of accident location, the mobile numbers are fed with the help of software programming of microcontroller. The various components used in this model are given in Table 4.1.

A. **Arduino UNO**

The ATmega328P microcontroller is used on the Arduino UNO for controlling the complete process. This acts as the heart of the device that can be used to control up to six PWM outputs on 14 digital I/O ports. Figure 4.2 depicts various components of the Arduino UNO. We need to start microcontroller by including a USB card and alternating current to direct current (DC) converter or battery. The Arduino boards are easy to use and do not require a high level of knowledge to operate them [11,12].

The board acts as the core of the microcontroller. The microcontroller, on the other hand, is a single chip that contains all the necessary components, such as CPU, RAM, and flash memory. To program the Arduino UNO board, we have used programming codes. The digital pins of ATmega328 0 (R_X) and 1 (T_X) can be used for serial (T_X) connection of UART TTL (5 V). The on-board ATmega16U2 serves as a virtual com port for PC software, which allows serial communication via USB. To avoid irreversible damage to the microcontroller, a 40 mA I/O pin should be used. For each analog input, UNO supports an accuracy of 10 bits. They are named from A0 to A5 (i.e., 1024 different values). There is an AREF function for the analog pin and reference, which can be used to adjust the upper end of the measurement range from 0 to 5 V [11,15].

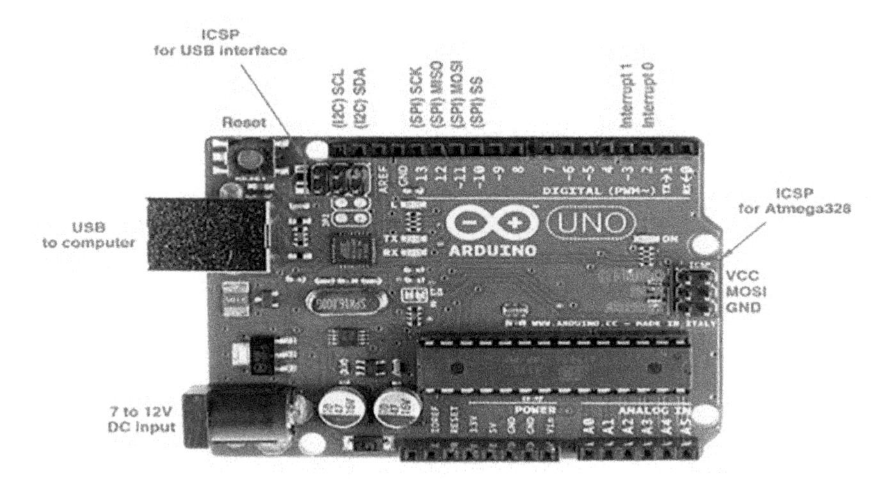

Figure 4.2 Arduino UNO [12–14].

B. **Accelerometer Sensor**

To detect crash or severe collision, the accelerometer sensor is used in the model [15,16]. The dynamic range of ADXL335 model measures the maximum amplitude of distortion, which is evaluated in g's (gyro). This sensor will send a signal to the controller, which will activate the GPS and GSM modules to send message to the pre-programmed numbers. The accelerometer sensor is shown in Figure 4.3.

For the accelerometer to work, mechanical energy must be converted into electrical energy. When the weight is placed on the sensor, which acts like a spring, it begins to descend. When it starts to land, it starts to accelerate. An accelerometer sensor is a device that monitors the acceleration of an object or object at its rest. Many electronic gadgets, cell phones, and other wearable technologies include accelerometer sensors.

To measure changes in the device position, acceleration is translated as an electrical signal. On analog and digital devices, the accelerometer can be found. Capacitance and piezoelectric sensors are two terms most commonly used to describe how this tool works, but it is really a complex electronic circuit made up of many different components, each with its own purpose and operating in a different way. Acceleration forces cause small crystal structures to tighten. These pressures generate voltages, which is used by the accelerometer to measure speed and direction. The piezoelectric effect is the most common type of accelerometer [15]. There are many industrial and scientific applications for accelerometers. Aircraft and missile inertial navigation systems use highly sensitive accelerometers. Accelerometers are used to track the frequency and magnitude of vibration in moving parts. As a result, photographs on tablets and digital cameras are always displayed vertically. Accelerometers are used to keep the aircraft in the air. Proof mass displacement is not immediately perceived in these accelerometers. As

Figure 4.3 Accelerometer sensor [12,14].

a result, the proof block is mechanically attached to the sensor, like a beam, as the proof block moves, subject to sensor force. The power changes the equivalent rigidity of the sensor, changing the resonant frequency, which is usually in the range of 100–500 kHz. An effective way to overcome the limitations of non-resonant accelerometers is to combine a large mass of guide mass with a high-quality sensor.

C. **Global Positioning System**

The transmitter and receiver pins of GPS module are directly interfaced with Arduino and a serial communication of data transfer is made available for the system. The GPS module is used to detect the current coordinates, which includes latitudes as well as longitudes. When receiving an incoming call, it is possible to see that the ring is written on the serial screen. Depending on the execution environment, the coverage area of each cell varies. Antennas used in large cells are mounted on masts or buildings taller than a normal roof. Microcell antenna heights are generally lower than the average city surface height, making them ideal for deploying in urban areas. Typically, Pico cells are used indoors because their coverage area is limited to a few tens of meters.

To connect to the network via broadband, Femto Cells are intended for home or small business use [17]. Small cells with shaded areas and gaps in coverage can benefit from canopy cells. Contingent upon the circumstance, the sign can go from a couple of hundred meters to many kilometers or more. Involving GSM up to a distance of 35 km in practice is conceivable. It is likewise conceivable to expand the possibility of the cell to a bigger range, and this should be possible in various ways relying upon the receiving wire framework, the territory type and the time passed beginning to end. GPS is depicted in Figure 4.4.

Figure 4.4 GPS module [14,18].

Broadband Internet associations are utilized to interface the correspondence supplier organization to photocells, cells expected for use in private and independent company conditions. Little cells with concealed regions and holes in inclusion can profit from shade cells. All parts of the GSM organization—cell phone, base station subsystem, network switch subsystem, working framework, and backing consistent help—are interconnected. The SIM card provides data about the cell phone client, for example, their name and telephone number, to the organization. Base station subsystem goes about as a correspondence channel between the network switch subsystem and the cell phone. The base handset station and the base station control (BSC) unit are two fundamental parts. The remote transmitters, collectors, and radio wires in the base handset station are associated with the cell phones, while the cerebrum of the framework is in the BSC. The base handset stations are managed and informed by the BSC.

D. **GSM Module**

The transmitter and receiver pins of the GSM module are directly interfaced with the Arduino UNO. The SIM800L modem was created utilizing SIMCOM's double band GSM/GPRS-based SIM800L modem. GSM is shown in Figure 4.5. It works on frequencies of 800/1800 MHz. SIM800L can look through the two groups naturally. Recurrence groups can likewise be tuned through AT orders. Awful rate can be changed from 1200 to 115,200 utilizing AT order. The GSM/GPRS modem has an implicit TCP/IP stack to permit you to interface with the Internet through GPRS [16,19,20]. The SIM800L is a tiny and dependable remote module. It is based on the SMT type with an across-the-board GSM/GPRS module and the most remarkable single-chip CPU.

Microcell antenna heights are generally lower than the average city surface height, making them ideal for deploying in urban areas. Typically, Pico cells are used indoors because their coverage area is limited to a few tens of meters. To connect to the network via broadband, Femto Cells are intended for home or small business use. Small

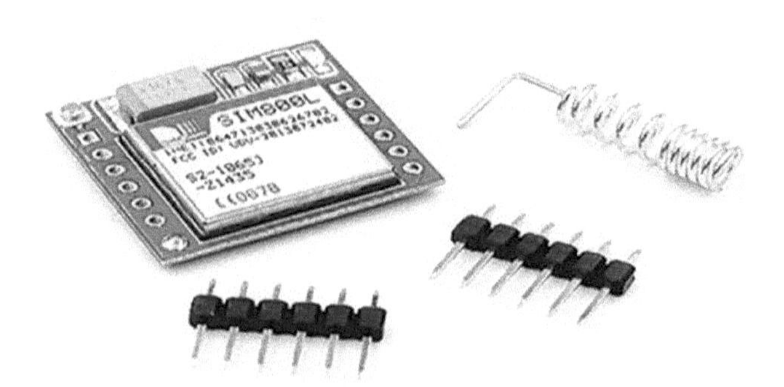

Figure 4.5 GSM module [14,18].

cells with shaded areas and gaps in coverage can benefit from canopy cells. Contingent upon the circumstance, the sign can go from two or three hundred meters to many kilometers or more. Involving GSM up to a distance of 35 km in practice is conceivable. It is likewise conceivable to expand the possibility of the cell to a bigger range, and this should be possible in various ways relying upon the receiving wire framework, the landscape type, and the time passed beginning to end. Inclusion can be given in the GSM building, either with an indoor Pico-cell base station or with a repeater framework with inward dispersed receiving wires provided by power splitters [21–23].

E. **Gear Motor**

A geared motor is an electric motor coupled with a gearbox whose mechanism adjusts the speed of the motor. In this model, geared motor is used to drive the prototype model. Like the torque coefficient, gear heads allow smaller motors to operate at higher speeds, allowing gear motors to deliver higher torque at lower speeds. Gear motors use either alternating current or DC. Often, the gearbox is used to reduce motor shaft speed and increase torque output efficiency.

It is advantageous for the gears to have fewer pieces if the driver and followers are close to each other. Since gears require more lubrication, the hourly maintenance cost may be higher. Typical DC motors consist of a stator with a magnet stationary and a motor with a soft iron core and one or more coils of insulated wire wrapped around it. Larger motors may have multiple parallel current paths due to multiple turns of the windings around the core. The commutator wire connects the ends of the coil. The brush connects the rotating coils to the external power supply through the commutator, allowing the motor coils to operate individually. The external view of the gear motor is shown in Figure 4.6.

Figure 4.6 Gear motor [14].

4.4 RESULT AND DISCUSSION

The experimental model was made according to the circuit diagram and the results were acquired as expected. At the point, when accident of the vehicle is occurring then the gadget will deliver a message to the family members' mobile, which will reflect the following message on the given cell phone.

1."Accident alert Latitude: 2400.0090, N Longitude: 12100.0000, E time: 12:00"	2."Accident alert Latitude: 25.44, N Longitude: 81.75, E time: 15:00"

The proposed model shows the exact location or area of the motor vehicle on the mobile number provided and also on the LCD screen mounted on it which likewise ensures the proper functioning state of the microcontroller circuit of the device.

Figure 4.7 clearly shows all four parts of working IoT-based vehicle model for GPS activation, accident alert, and display of latitude and longitude.

4.5 CONCLUSION

The response of the emergency services is one of the significant factors during vehicles collision, which leads to severe accidents becoming fatal. In maximum cases, these are not intimated within enough time as the vehicle's passenger may be unable to make calls/raise alarms. Therefore, the automation of this concept provides more real, competent and exact data, which will not waste the valuable time of emergency services. The main goal of the IoT-based Accident Alert and Vehicle Tracking System is to reduce the death rate from accidents. Whenever an alert of the accident is raised, the paramedics arrive at the specific site to enhance the probability

Figure 4.7 IoT-based vehicle model for GPS activation and display of latitude and longitude.

of survival. The tracking and accident alert features of the vehicle are much more helpful for accidents in deserted places and on midnights. The main objective of this chapter is to decrease the death rate from severe accidents and provide assistance to the victim within the time to increase the chances of survival. Our system will detect the incidence of a fatal accident and intimate the emergency services about the correct location of this collision and a selected number of emergency contacts will be informed. We need to send a message to the GSM device for tracking the vehicle, for which it gets activated. The shock sensor connected to the vehicles can also activate the GSM device by detecting accidents. The buffer's last latitude and longitude position values are taken by the GSM. It sends a message to the predefined specific contact number or laptop in the program. After the message is propagated to the predefined device, the GSM is deactivated, and GPS is activated. Some disadvantages may include natural disaster circumstances, which may load the network with more data than the handling capacity of the network. Also, if the navigation services are down or there is no network connectivity, the system will be unable to notice the site due to a hindrance in the device's operation. Fleet monitoring, driver monitoring, route monitoring, vehicle scheduling and accident analysis are some of the applications of this concept.

REFERENCES

[1] M. Balfaqih, S. A. Alharbi, M. Alzain, F. Alqurashi, and S. Almilad, "An Accident Detection and Classification System Using Internet of Things and Machine Learning towards Smart City", *Sustainability*, vol. 14, no. 1, 210, 2021.

[2] M. Fogué, P. Garrido, F. J. Martinez, J.-C. Cano, C. T. Calafate, and P. Manzoni, "Automatic Accident Detection: Assistance Through Communication Technologies and Vehicles", *IEEE Vehicular Technology Magazine*, vol. 7, no. 3, pp. 90–100, Sept. 2012, doi: 10.1109/MVT.2012.2203877.

[3] U. Alvi, M. A. K. Khattak, B. Shabir, A. W. Malik, and S. R. Muhammad, "A Comprehensive Study on IoT Based Accident Detection Systems for Smart Vehicles," *IEEE Access*, vol. 8, pp. 122480–122497, 2020.

[4] J. Mounika, N. Charanjit, B. Saitharun, and B. Vashista, "Accident Alert and Vehicle Tracking System using GPS and GSM (April 17, 2021)," *Asian Journal of Applied Science and Technology (AJAST)*, vol. 5, no. 2, pp. 81–89, April-June 2021, Available at SSRN: https://ssrn.com/abstract=3869132.

[5] Sawant, Kiran, Imran Bhole, Prashant Kokane, Piraji Doiphode, and Yogesh Thorat, "Accident Alert and Vehicle Tracking System," *International Journal of Innovative Research in Computer and Communication Engineering*, vol. 4, no. 5, pp. 8619–8623, 2016.

[6] S. R. Aishwarya, A. Rai, Charitha, M. A. Prasanth, and S. C. Savitha, "An IoT Based Vehicle Accident Prevention and Tracking System for Night Drivers", *International Journal of Innovative Research in Computer and Communication Engineering*, vol. 3, no. 4, pp. 2320–9798, April 2015.

[7] B. Shabrin, M. Poojary, T. Pooja, and B. Sadhana, "Smart helmet-intelligent safety for motorcyclist using Raspberry Pi and Open CV, *International Research Journal of Engineering and Technology (IRJET),* 3, no. 3 (2016): 589–593.

[8] M. Sheth, A. Trivedi, K. Suchak, K. Parmar, and D. Jetpariya, "Inventive Fire Detection utilizing Raspberry Pi for New Age Home of Smart Cities, " *2020 Third International Conference on Smart Systems and Inventive Technology (ICSSIT), 2020*, pp. 724–728, doi: 10.1109/ICSSIT48917.2020.9214108.

[9] E. O. Elamin, W. M. Alawad, E. Yahya, A. Abdeen, and Y. M. Alkasim. "Design of Vehicle Tracking System", In *2018 International Conference on Computer, Control, Electrical, and Electronics Engineering (ICCCEEE)*, pp. 1–6, 2018.

[10] A. H. Alquhali, M. Roslee, M. Y. Alias, and K. S. Mohamed, "Iot based real-time vehicle tracking system", In *2019 IEEE Conference on Sustainable Utilization and Development in Engineering and Technologies (CSUDET)*, pp. 265–270, 2019.

[11] I. Rajput et al., "Attendance Management System using Facial Recognition," *3rd International Conference on Intelligent Engineering and Management (ICIEM)*, 2022, pp. 797–801, doi: 10.1109/ICIEM54221.2022.9853048.

[12] M. Varanis, A. Silva, A. Mereles, and Robson Pederiva, "MEMS Accelerometers for Mechanical Vibrations Analysis: A Comprehensive Review with Applications." Journal of the Brazilian Society of Mechanical Sciences and Engineering, vol. 40, p. 527, 2018. https://doi.org/10.1007/s40430-018-1445-5.

[13] S. Ferdoush, and X. Li, "Wireless Sensor Network System Design Using Raspberry Pi and Arduino for Environmental Monitoring Applications," Procedia Computer Science, vol. 34, pp. 103–110, 2014. https://doi.org/10.1016/j.procs.2014.07.059.

[14] https://create.arduino.cc/projecthub/electropeak/the-beginner-s-guide-to-control-motors-by-arduino-and-l293d-139307. (https://projecthub.arduino.cc/samanfern/bluetooth-controlled-car-c71cd0).

[15] T. Anitha, and T. Uppalaiah, "Android Based Home Automation Using Raspberry pi," International Journal of Innovative Technologies, vol. 4, no. 01, pp. 1154–1156, 2016.

[16] N. Watthanawisuth, T. Lomas, and A. Tuantranont, "Wireless black box using MEMS accelerometer and GPS tracking for accidental monitoring of vehicles," *Proceedings of 2012 IEEE-EMBS International Conference on Biomedical and Health Informatics*, pp. 847–850, 2012, doi: 10.1109/BHI.2012.6211718.

[17] S. Bhavthankar, and H. G. Sayyed. "Wireless System for Vehicle Accident Detection and Reporting Using Accelerometer and GPS," *International Journal of Scientific & Engineering Research*, vol. 6, no. 8, pp. 1068–1071, 2015.

[18] P. Yellamma, N. S. N. S. P. Chandra, P. Sukhesh, P. Shrunith and S. S. Teja, "Arduino Based Vehicle Accident Alert System Using GPS, GSM and MEMS Accelerometer," 2021 5th International Conference on Computing Methodologies and Communication (ICCMC), 2021, pp. 486–491, doi: 10.1109/ICCMC51019.2021.9418317.

[19] I. Guvenc, S. Saunders, O. Oyman, H. Claussen, and A. Gatherer, "Femtocell Networks", *Journal on Wireless Communications and Networking,* pp. 1–2, 2010, https://doi.org/10.1155/2010/367878.
[20] A. Anusha, and S. M. Ahmed, "Vehicle Tracking and Monitoring System to Enhance the Safety and Security Driving Using IoT," 2017 International Conference on Recent Trends in Electrical, Electronics and Computing Technologies (ICRTEECT), pp. 49–53, 2017, doi: 10.1109/ICRTEECT.2017.35.
[21] M. S. Mahmud, H. Wang, A. M. Esfar-E-Alam, and H. Fang, "A Wireless Health Monitoring System Using Mobile Phone Accessories," *IEEE Internet of Things Journal*, vol. 4, no. 6, pp. 2009–2018, Dec. 2017, doi: 10.1109/JIOT.2016.2645125.
[22] S. Hong, and D. Park, "Lightweight Collaboration of Detecting and Tracking Algorithm in Low-Power Embedded Systems for Forward Collision Warning," *Twelfth International Conference on Ubiquitous and Future Networks (ICUFN)*, pp. 159–162, 2021, doi: 10.1109/ICUFN49451.2021.9528771.
[23] M. N. Ramadan, M. A. Al-Khedher, and S. A. Al-Kheder. "Intelligent Anti-Theft and Tracking System for Automobiles," *International Journal of Machine Learning and Computing*, vol. 2, no. 1, p. 83, 2012, doi:10.7763/IJMLC. 2012.V2.94.

Chapter 5

Intelligent control for passenger safety

Overview, dynamic modeling and control of active suspension systems

Himanshu Chhabra, Urvashi Chauhan, and Prince Jain
Parul University

Bhavnesh Kumar
Netaji Subhas University of Technology

CONTENTS

DOI: 10.1201/9781003436089-5

ABBREVIATIONS

DOF	Degree of Freedom
LFT	Linear Fractional Transformations
PID	Proportional Integral Derivative
QVASS	Quarter Vehicle Active Suspension System

5.1 INTRODUCTION: OVERVIEW AND REQUIREMENT OF SUSPENSION SYSTEM IN VEHICLE

A vital part of a vehicle's system that ensures the passengers' comfort and safety is the suspension system. A suspension system's three primary parts are the spring, linkage, and shock absorber. In suspension systems, the energy is stored via spring and dissipates through a damper. Suspension system reduces the effect of body acceleration and vertical displacement caused due to uncertainty and road disturbances. The performance of a vehicle, i.e., safety ride and road handling capability, is determined by its suspension system, which provides the required forces between the vehicle and the road. To achieve comfort ride, the soft suspension system should be needed whereas stiff suspension is required to improve the road handling capability [1,2]. These two conflicting objectives need to be possessed simultaneously in order to achieve the desired vehicle performance. Thus, the designing and controlling of suspension systems is quite challenging. Various researchers have gained remarkable attention in this field [3–6]. The quarter car active suspension model is thoroughly covered in this chapter. Furthermore, the robust control strategies for an active suspension system have been presented. H-infinity control and μ-Synthesis Control techniques are effectively analyzed for dynamic model of a quarter car active suspension system.

5.2 CHALLENGES IN DESIGNING THE VEHICLE SUSPENSION CONTROL SYSTEM

In general, the most difficult task in designing a vehicle controller is modeling the behavior of the plant. There are a variety of reasons that cause difficulty in designing optimum controller such as improper system modeling, time-varying nature of system variable, nonlinear, and complex system dynamics. The major issue in designing the suspension system is the difficulty in simultaneous occurring of two objectives that are conflicting each other. Both the objectives need to be accomplished simultaneously for the efficient performance of vehicle. If any of one objective is fail, the model does not meet the desired requirement thus found ineffective as the purpose is to provide the ride comfort to the passengers as well as roads holding capabilities [7–9]. For the comfort ride, the soft suspension is required whereas to control the load handling stiff suspension is required. Thus, to design a suspension system that has the aforementioned requirements is quite difficult. This area has gained the remarkable attention as human safety is the primary requirement.

5.3 CHARACTERIZATION OF SUSPENSION SYSTEMS

Three types of suspension systems are well known in transportation, i.e., passive suspension system, semi-active suspension system, and active suspension system. The following provides a detailed description of each type of suspension system.

5.3.1 Passive suspension system

Passive suspension system is known to have a simple mechanism and easy implementation. Moreover, these systems are reliable and have inexpensive design requirements. However, the system is inadequate in achieving ride comfort and roads holding objectives simultaneously due to the constant spring and damper characteristics. The invariable system parameters are unable to deal with the random road disturbances and conflicting objectives thus not to be preferred in recent scenarios [10]. These problems incurred can be effectively coped by means of a semi-active and active suspension system. Figure 5.1 illustrates the passive suspension system, which is a typical mechanical components arrangement

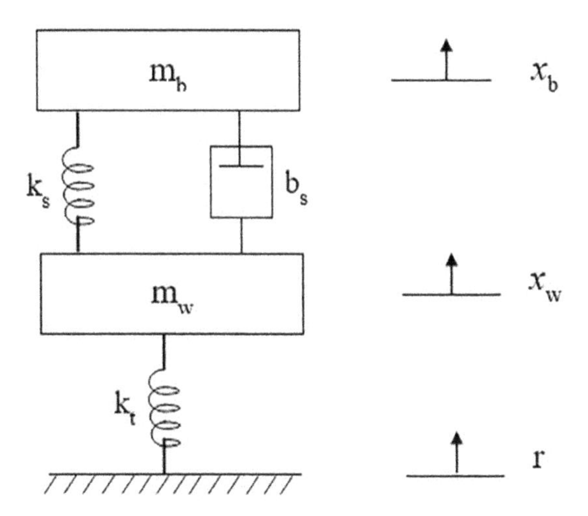

Figure 5.1 Passive suspension system.

comprising damper and spring that have constant parameters, i.e., damping coefficient b_s and stiffness k_s, respectively. m_b represents the sprung mass whereas m_w, is the unsprung mass (i.e., axle and wheel assembly's mass). k_t is the coefficient of tire compressibility. The input of the road disturbance acting on the unsprung mass is denoted by r and vertical displacements of the sprung mass and the unsprung mass are denoted by x_b and x_w, respectively.

Damper is equipped with either hydraulic fluid or compressed gas in it that is allowed to move through a hole using a piston. A force that is proportional to the speed differential between the unsprung and sprung masses is produced by fluid motion. The passive suspension system is insufficient to achieving the requirements and cannot yield decent outcomes as there is no external control provided.

5.3.2 Semi-active suspension system

Semi-active suspension models possess varying damper coefficients, thus providing a safe ride in contrast with a passive suspension system. However, the spring coefficient is still constant in the case of a semi-active suspension system. Also, the resulting forces are limited by margin constraints that bound the comfort riding. With these systems, it is possible to switch seamlessly between a passive damper with a semi-active damping coefficient and a passive damper. The continuously variable damper employed alters the capacity in real time using closed feedback loop control design, improving overall suspension performance in comparison to a passive system. This allows for optimal energy dissipation while preserving fixed spring utilization. To stiffen or soften the system, the damping coefficient of a semi-active suspension can be altered intermittently or continuously [11,12]. By minimizing the low-frequency response to inertial forces, new operating tactics, such as strengthening the suspension, would improve the vehicle's cornering, braking, and accelerating.

Semi-active suspension is categorized on damper characteristics (Figure 5.2).

- Orifice-based damper
- Magnetorheological fluid-based damper

5.3.3 Active suspension system

Active suspension system has a wide range of applications as it overcomes the shortcomings of the two aforesaid suspension models as it has the ability to generate independent forces of relative suspension motion. Nevertheless, the proper designing mechanism and selection of

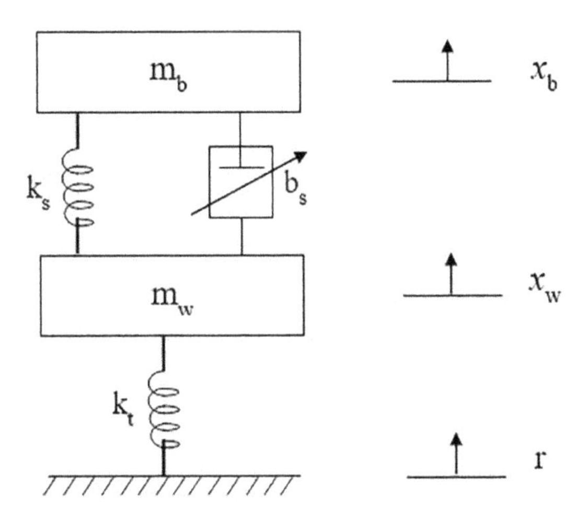

Figure 5.2 Semi-active suspension system.

an appropriate control algorithm is quite essential and also challenging due to the adversity of road conditions and other criteria. In 1958, the idea of an active suspension system was first presented. One of the key research areas in the modern car industry is an active suspension system with additional control force to suppress the oscillations because the existing passive and semi-active suspension systems' ability to reduce vibration is constrained. The active suspension system differs from conventional suspension in that it can inject energy into the dynamics of the vehicle through actuators rather than waste it. Actuators are positioned between the unsprung mass and the sprung mass in active suspension systems to provide control force to respond in real time to a range of road disturbances. Enhancing ride comfort by absorbing shocks from a rough and uneven road is the main goal of building an active suspension system [13–16]. The force actuator in the active suspension system is capable of adding and releasing the system's energy. As a result, the suspension system gains the capacity to manage the vehicle's posture, lessen the impact of braking, and reduce vehicle roll during cornering and braking in addition to improving ride comfort and handling.

Figure 5.3 shows the active suspension system, where B_s is the damping coefficient and k_s denotes the stiffness coefficient. m_s and m_u represent the sprung mass and unsprung mass, respectively. k_t is the coefficient of tire compressibility and f_A Stands for the control force produced by the actuator. The road disturbance input acting on the unsprung mass is denoted by r and vertical displacements of the sprung mass and the unsprung mass are denoted by x_b and x_w, respectively.

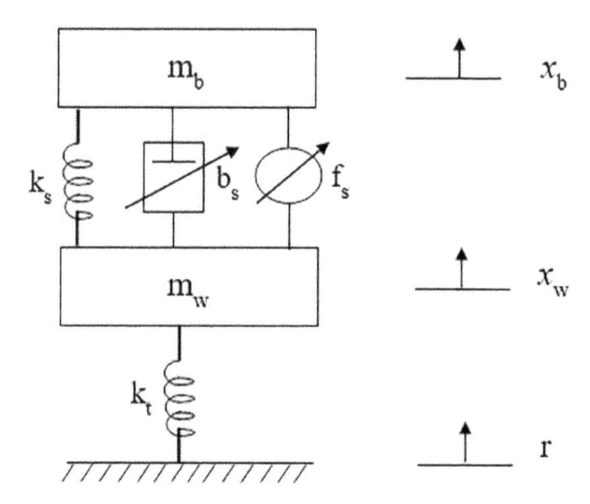

Figure 5.3 Active suspension system.

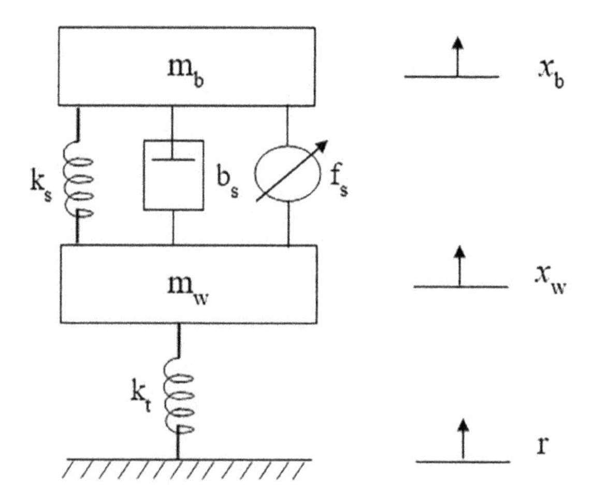

Figure 5.4 Mechanical model of active suspension system.

5.4 TWO-DEGREE OF FREEDOM (2-DOF) QUARTER CAR ACTIVE SUSPENSION SYSTEM MODEL

Figure 5.4 shows a quarter vehicle model for an active suspension system. The parameter values considered are specified in Table 5.1.

The mass of the vehicle's body, frame, and interior components that the suspension supports is represented by the sprung mass, or m_b. The axle and wheel assembly's mass, m_w, is the unsprung mass; k_s and b_s are the spring and damper coefficients of the suspension system's passive parts, respectively. Tire

Table 5.1 Model parameters of active suspension system (Singh, Chhabra, and Bhangal 2016)

Model parameters	*Parameter*	*Specification*
Vehicle body mass	m_b	300 kg
Wheel assembly mass	m_w	60 kg
Suspension stiffness	k_s	16,000 N/m
Suspension damping	b_s	1000 N s/m
Tire stiffness	k_t	1,90,000 N/m
Force	f_s	Actuator force
Disturbance	R	Road disturbance

compressibility is the coefficient k_t. f_s stands for the control force produced by the actuator, and r stands for the input of the road disturbance acting on the unsprung mass. The vertical displacements of the sprung mass and the unsprung mass are denoted by the symbols x_b and x_w, respectively [17].

5.4.1 Mathematical modeling

The mathematical equation of motion that depicts the system dynamics is written using Newton's second principle and free body diagram approach, given as follows:

Equation for sprung mass:

$$m_b \ddot{x}_b = k_s\left(x_w - x_b\right) + b_s\left(\dot{x}_w - \dot{x}_b\right) + f_s \tag{5.1}$$

Equation for unsprung mass:

$$m_w \ddot{x}_w = k_t\left(r - x_w\right) - k_s\left(x_w - x_b\right) - b_s\left(\dot{x}_w - \dot{x}_b\right) - f_s \tag{5.2}$$

5.4.2 State space model of an active suspension system

Now the dynamic model of the suspension system (Equations (5.1) and (5.2)) is expressed into the state variable form. Let the state $x = \left[\ x_1 x_2 x_3 x_4\right]^T$ be defined as

$$x_1 = x_b \tag{5.3}$$

$$x_2 = x_w \tag{5.4}$$

$$x_3 = \dot{x}_b \tag{5.5}$$

$$x_4 = \dot{x}_w \tag{5.6}$$

The dynamic model of the suspension system described in Equations (5.1)–(5.6) is now expressed into the state equation as follows:

$$\dot{x}_1 = x_3 \tag{5.7}$$

$$\dot{x}_2 = x_4 \tag{5.8}$$

$$\dot{x}_3 = \frac{1}{m_b}\left(k_s\left(x_2 - x_1\right) + b_s\left(x_4 - x_3\right) + f_s\right) \tag{5.9}$$

$$\dot{x}_4 = \frac{1}{m_w}\left[k_t\left(r - x_2\right) - k_s\left(x_2 - x_1\right) - b_s\left(x_4 - x_3\right) - f_s\right] \tag{5.10}$$

5.4.2.1 Uncertain modeling

$$k_s = \bar{k}_s\left(1 + \delta_k P_k\right) \tag{5.11}$$

$$b_s = \bar{b}_s\left(1 + \delta_b P_b\right) \tag{5.12}$$

Where the nominal values of the associated spring constant and damping coefficient are $\bar{k}_s$ & $\bar{b}_s$, respectively. In this work, the maximum relative uncertainty of ±10% in spring constant and ±20% in damping coefficient has been considered. Thus, the parameters P_b and P_k have been considered as 0.1 and 0.2, respectively. δ_b and δ_k range as $\left(-1 \le \delta_b, \delta_k \le 1\right)$.

In Figure 5.5, f_s shows the control force produced by the actuator and the road disturbance input acting on the unsprung mass is denoted by r. V_b and V_k represent the perturbation in damper and spring, respectively, whereas u_b and u_k are the uncertainty in damper and spring input, respectively.

Unstructured uncertainty is expressed as a multiplicative perturbation added to the actuator's input as shown in Figure 5.6, where W_m shows the unstructured uncertainty added in actuator input and $\bar{G}_{act}$ denotes the actuator model without uncertainty. Figure 5.7 shows the overall system model including uncertainty in system parameters and actuator input, and Figure 5.8 shows the input–output relationship. An upper linear fractional transformation (LFT) representation in Figure 5.7 can be used to describe the overall suspension system's unpredictable behavior as follows:

$$G_{susp} = \left[\begin{array}{c|cc} [A]_{4\times4} & [B_1]_{4\times4} & [B_2]_{4\times2} \\ \hline [C_1]_{4\times4} & [D_{11}]_{4\times4} & [D_{12}]_{4\times2} \\ [C_2]_{4\times4} & [D_{21}]_{4\times4} & [D_{22}]_{4\times2} \end{array}\right] \tag{5.13}$$

$$\Delta_{susp} = \begin{bmatrix} \Delta_b & 0 \\ 0 & \Delta_k \end{bmatrix} \tag{5.14}$$

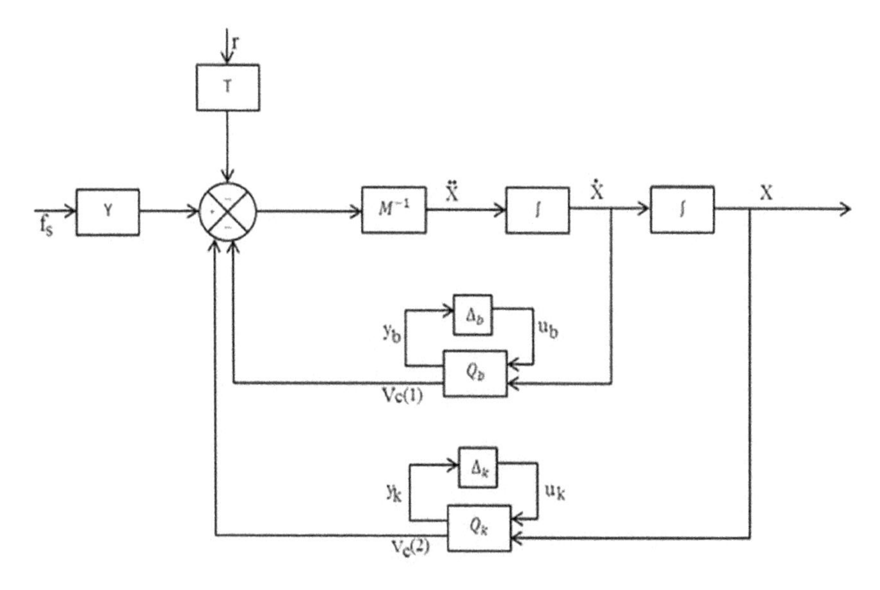

Figure 5.5 System block diagram with uncertain parameters.

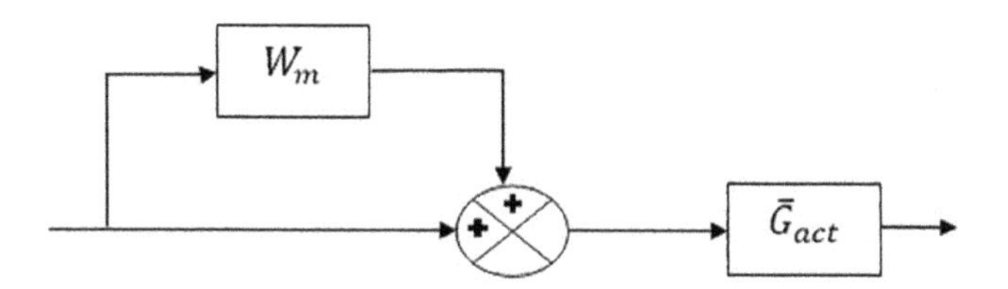

Figure 5.6 Uncertainty in actuator model.

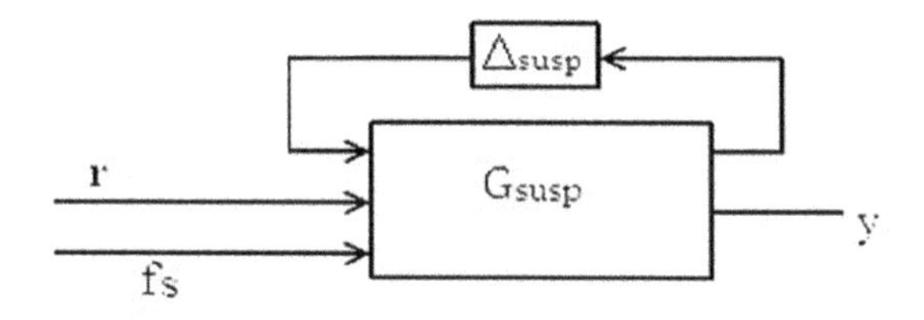

Figure 5.7 Block diagram of system with uncertainty.

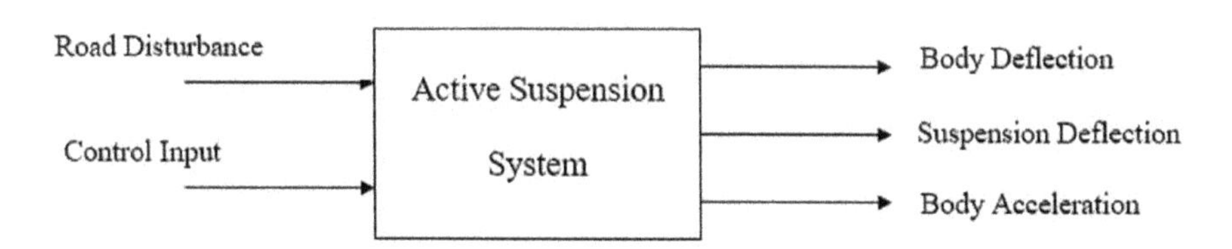

Figure 5.8 General block diagram.

The following are the state equation and output equation:

$$\begin{bmatrix} \dot{x}_1 \\ \dot{x}_2 \\ \dot{x}_3 \\ \dot{x}_4 \end{bmatrix} = \begin{bmatrix} 0 & 0 & 1 & 0 \\ 0 & 0 & 0 & 1 \\ -53.3 & 53.3 & -3.3 & 3.3 \\ 266.7 & -3433.3 & 16.7 & -16.7 \end{bmatrix} \begin{bmatrix} x_1 \\ x_2 \\ x_3 \\ x_4 \end{bmatrix} +$$

$$\begin{bmatrix} 0 & 0 & 0 & 0 & \vdots & 0 & 0 \\ 0 & 0 & 0 & 0 & \vdots & 0 & 0 \\ -0.3 & 0.3 & -10.7 & 10.7 & \vdots & 0 & 33.3 \\ 1.7 & -1.7 & 53.3 & -53.3 & \vdots & 3166.7 & -166.7 \end{bmatrix} \begin{bmatrix} u_b \\ u_k \\ r \\ f_s \end{bmatrix} \tag{5.15}$$

$$\begin{bmatrix} y_b \\ y_k \\ x_b \\ s_d \\ a_b \end{bmatrix} = \begin{bmatrix} 0 & 0 & 1 & 0 \\ 0 & 0 & 0 & 1 \\ 1 & 0 & 0 & 0 \\ 0 & 1 & 0 & 0 \\ \cdots & \cdots & \cdots & \cdots \\ 1 & 0 & 0 & 0 \\ 1 & -1 & 0 & 0 \\ -53.3 & 53.3 & -3.3 & 3.3 \end{bmatrix} \begin{bmatrix} x_1 \\ x_2 \\ x_3 \\ x_4 \end{bmatrix} + \tag{5.16}$$

$$\begin{bmatrix} 0 & 0 & 0 & 0 & \vdots & 0 & 0 \\ 0 & 0 & 0 & 0 & \vdots & 0 & 0 \\ 0 & 0 & 0 & 0 & \vdots & 0 & 0 \\ 0 & 0 & 0 & 0 & \vdots & 0 & 0 \\ \cdots & \cdots & \cdots & \cdots & \cdots & \cdots & \cdots \\ 0 & 0 & 0 & 0 & \vdots & 0 & 0 \\ 0 & 0 & 0 & 0 & \vdots & 0 & 0 \\ 0 & 0 & 0 & 0 & \vdots & 0 & 33.3 \end{bmatrix} \begin{bmatrix} u_b \\ u_k \\ r \\ f_s \end{bmatrix}$$

5.5 INTELLIGENT CONTROL TECHNIQUES FOR ACTIVE SUSPENSION SYSTEM

In the past few years, different types of controllers are used in robotic control and various industries for several purposes. In order to incorporate knowledge of how to best control a system, a modern approach of robust controller plays an important role. In this chapter, the theoretical background robust control, H_∞ controller, and μ-synthesis controller are given.

5.5.1 H_∞ (i.e., "H-infinity") controller

In control theory, H_∞ techniques are used to design controllers that stabilize with guaranteed performance. By expressing the control problem as a mathematical optimization problem and then identifying the controller that resolves this optimization, a control designer can employ H_∞ techniques. Compared to traditional control methods, H_∞ approaches have the benefit of being easily adaptable to issues involving multivariate systems with cross-coupling between channels.

If a control system maintains stability and meets predetermined performance standards while facing potential uncertainties, it is considered robust. The Small Gain Theorem is crucial for carrying out stability testing using robust control methods.

Theorem 4.1

If $G_1(s)$ and $G_2(s)$ are stable, i.e., $G_1 \in H_\infty$, $G_2 \in H_\infty$, then the closed-loop system is internally stable if and only if

$$G_1G_{2\infty} < 1 \quad \textit{and} \quad G_2G_{1\infty} < 1 \tag{5.17}$$

A closed-loop system involving plant G and controller K is said to be robustly stable if it maintains stability across any disturbances on the plant that are technically achievable. Figure 5.9 shows the general close loop control with adding disturbance in the feedback path. We always assume that the perturbation set includes zero, it naturally implies that K is a stabilizing controller for the nominal plant G. Let's have a look at the example of additive perturbation shown in Figure 5.10, where $\Delta(s)$ is the perturbation and the "full" matrix is unknown but stable.

It is simple to determine that $T_{uv} = -K(I + GK)^{-1}$ is the transfer function from signal v to u. The controller K should stabilize the nominal plant G, as was already mentioned. As a result, we can derive the following conclusion from the Small Gain Theorem.

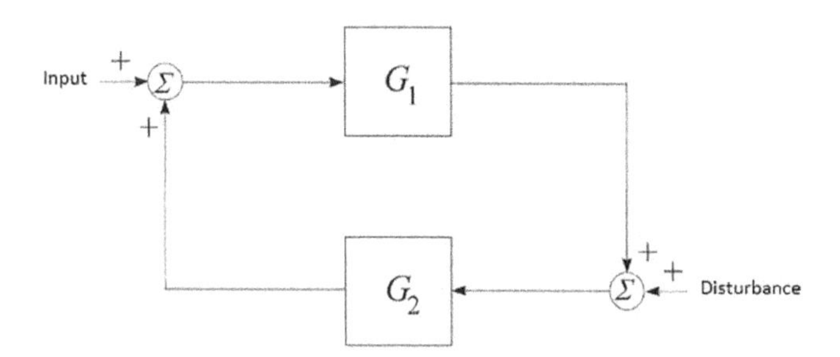

Figure 5.9 A feedback configuration.

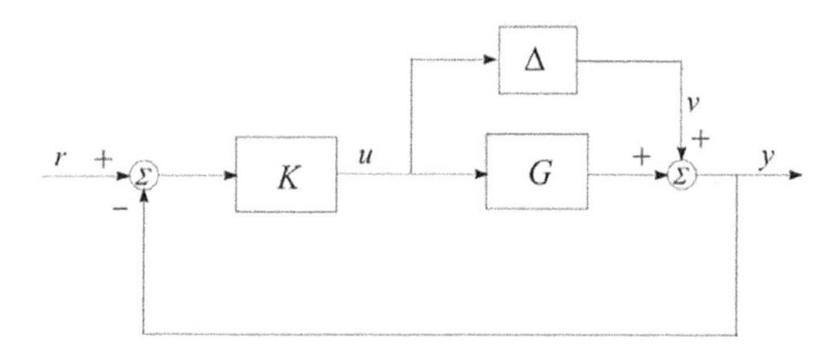

Figure 5.10 Additive perturbation configuration.

Theorem 4.2

For stable Δ(s), the closed-loop system is robustly stable if K(s) stabilizes the nominal plant and the following holds

$$\Delta K(I+GK)^{-1}{}_{\infty} < 1 \tag{5.18}$$

and

$$K(I+GK)^{-1}\Delta_{\infty} < 1 \tag{5.19}$$

or, in a strengthened form,

$$K(I+GK)^{-1}{}_{\infty} < \frac{1}{\Delta_{\infty}} \tag{5.20}$$

When the unknown Δ is capable of undergoing all phases, the second requirement becomes necessary. If establishing a controller that reliably stabilizes the widest range of disturbances, in the sense of the ∞-norm, is required, the following minimization issue must be solved.

$$\min_{K\ Stabilizing} K(I+GK)^{-1}{}_{\infty} \tag{5.21}$$

A sensible design would combine sensing, which calls for accurate tracking, with a restriction on the energy of the control signal. The following mixed sensitivity (or, so-called S over KS) problem may be of interest to us to solve (Figure 5.11).

$$\min_{K\ stabilizing} \begin{matrix} (I+GK)^{-1} \\ K(I+GK)^{-1} \end{matrix}{}_{\infty} \tag{5.22}$$

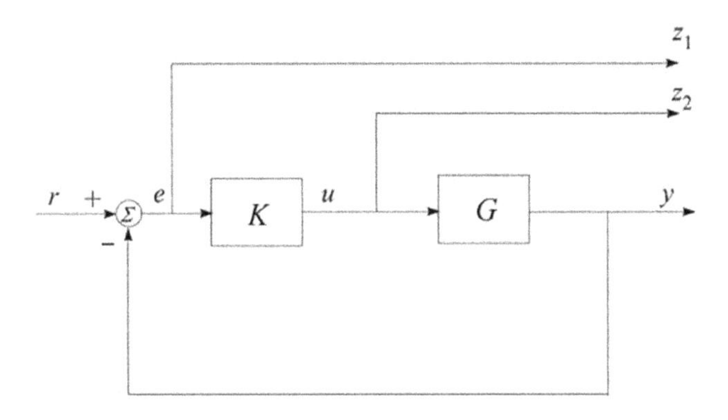

Figure 5.11 A mixed sensitivity consideration.

In terms of additive perturbation, this cost function can also be seen as the design goals of nominal performance (NP), good tracking, disturbance absorption, and durable stabilization.

It is challenging to simultaneously optimize robust performance (RP) and robust stabilization. H loop-shaping is one technique that comes close to accomplishing this; it enables the control designer to optimize the response near the system bandwidth to acquire strong robust stability (RS) and then use standard loop-shaping concepts to the multivariable frequency response to obtain good robust performance.

5.5.2 *μ*-synthesis controller

The open-loop interconnection system designated P includes all of the well-known components, such as the nominal plant model and the performance and uncertainty weighting factors. The uncertain component of the set Δ_{pert}, which parameterizes all of the problem's presumed model uncertainty, is the Δ_{pert} block. K is the controller. P is initially perturbed by three sets of inputs: disturbances d, controls u, and inputs z. There are three sets of outputs produced: measurements y, errors e, and perturbation outputs z.

The LFT (upper or lower) describes the group of systems that must be controlled:

$$\{F_U(P,\Delta_{pert}):\Delta_{pert},\max\sigma[\Delta_{pert}(j\omega)]\leq 1\} \quad (5.23)$$

The design goal is to identify a stabilizing controller K such that the closed-loop system is stable and satisfies for all such perturbations Δ_{pert}. A RP test on the linear fractional transformation $F_L(P, K)$ can be used to verify this performance objective for any K. When computing the RP test, an additional uncertainty structure should be taken into account.

$$\Delta := \left\{ \begin{bmatrix} \Delta_{pert} & 0 \\ 0 & \Delta_F \end{bmatrix} : \Delta_{pert} \in \Delta_{pert}, \Delta_F \in C^{n_d \times n_e} \right\} \tag{5.24}$$

The parametric uncertainties modeled in the QVASS are represented by the matrix's first block Δ, the uncertainty block Δ_{pert}. The second block, designated as Δ_F, is a fictional uncertainty block that was added to the μ-approach architecture to incorporate the performance targets. The structured singular value μ must satisfy the following requirement at each frequency in order for the stabilizing controller K to satisfy the design objectives:

$$\mu_{\Delta_P}\left[F_L(P,K)(j\omega)\right] < 1 \tag{5.25}$$

A suitable level of performance requirement can be established using normalization as

$$F_U(M,\Delta)_{\infty} < 1 \tag{5.26}$$

Equation (5.26) suggests that $F_U(M,\Delta)$ is stable, which signals RS in the face of plant disturbances Δ. The system loop in Figure 5.12 must satisfy Condition (Equation 5.26) in order to be robustly stable with respect to a made-up uncertainty block Δ_p. Δ_p is an unstructured performance uncertainty block that fulfills $\Delta_{p\infty} \leq 1$ and has the proper dimensions. Of course, $F_U(M,\Delta)$ is the portion enclosed by the dotted line.

Equation (5.26) ought to be true for all $\Delta \in B\Delta$ for reliable performance. According to Figure 5.12, a robust stabilization issue with the uncertainty block substituted by $\tilde{\Delta}$ can be used to describe both the RP design and robust stabilization against

$$\tilde{\Delta} \in \tilde{\Delta} := \{\text{diag}\{\Delta, \Delta_p\} : \Delta \in \boldsymbol{B}\Delta, \|\Delta_p\|_{\infty} \leq 1 \tag{5.27}$$

Accordingly, this is a RS problem for a structured uncertainty $\tilde{\Delta}$. Additionally, we have the following signs:

- RP $\leftrightarrow M_{\mu} < 1$ for structures uncertainty $\tilde{\Delta}$
- RS $\leftrightarrow M_{11\mu} < 1$ for structured uncertainty **B**Δ
- NP $\leftrightarrow M_{22\infty} < 1$
- Nominal stability $\leftrightarrow$ M is internally stable.

Naturally, the RS requirement corresponds to $M_{11\infty} < 1$ if the uncertainty is unstructured.

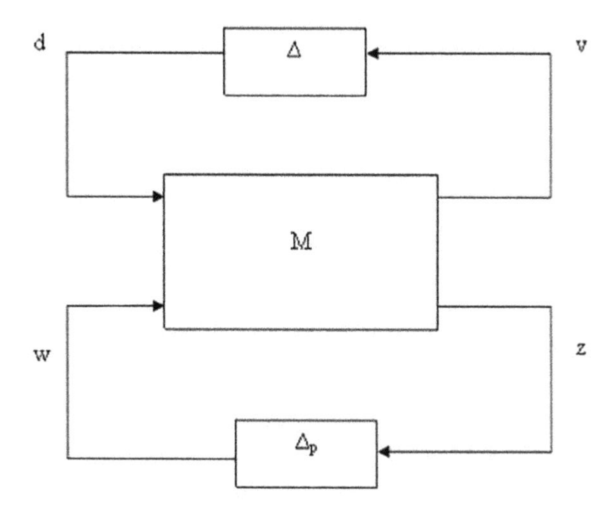

Figure 5.12 Standard M-Δ configuration with Δ_p analysis.

5.5.2.1 Disturbance rejection response using H-infinity and μ-synthesis controller

a. *Open-loop response*
The open-loop responses without controller for disturbance input (Equation 5.28) are shown in Figures 5.13–5.15. It shows that the Quarter Vehicle Active Suspension System is a highly unstable system. Therefore, the controller needs to be designed in order to stabilize the system and system effective against the disturbance input.

$$r(t) = \begin{cases} a\{1 - \cos(8 * pi * t)\} & 0 \le t \le 0.25 \\ 0 & \text{otherwise} \end{cases} \tag{5.28}$$

where $a=0.025$ (road bump height 5 cm).

b. *Response with H∞ controller*
H_∞ controller is designed for a Quarter Vehicle Active Suspension System. The control structure is implemented in MATLAB®. The disturbance rejection responses are shown in Figures 5.16–5.18.

So, H_∞ controller is successfully implemented for stabilization and control of the Quarter Vehicle Active Suspension System and its time domain specifications are analyzed. In the response of Body Acceleration, the values of settling time and peak overshoot are found to be 0.91 sec and 0.428 m/s^2, respectively. In the response of Suspension Deflection, the values of settling time and peak overshoot are found to be 1.1 sec and 0.00186 m, respectively. In the response

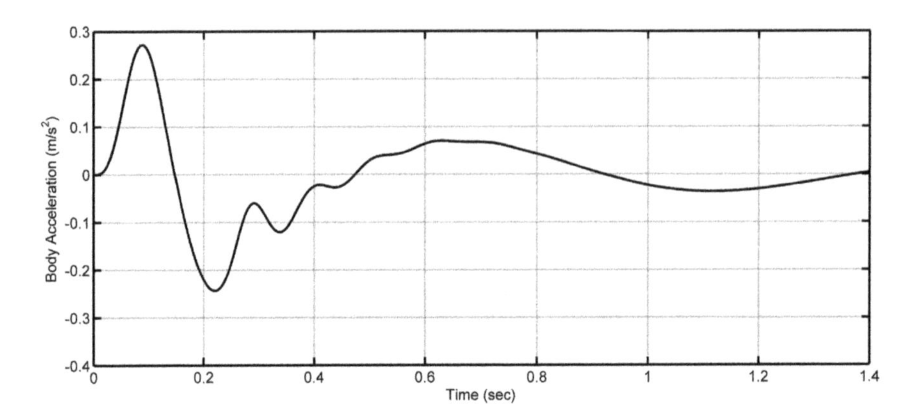

Figure 5.13 Open-loop response body acceleration.

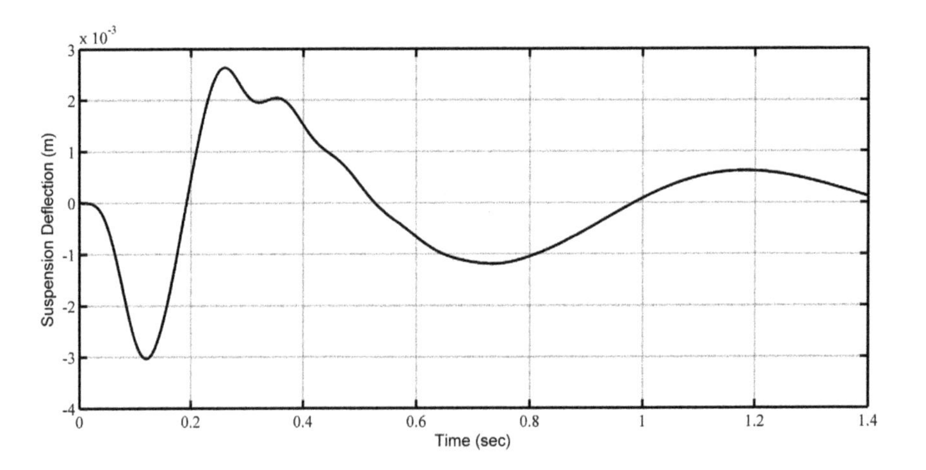

Figure 5.14 Open-loop response suspension deflection.

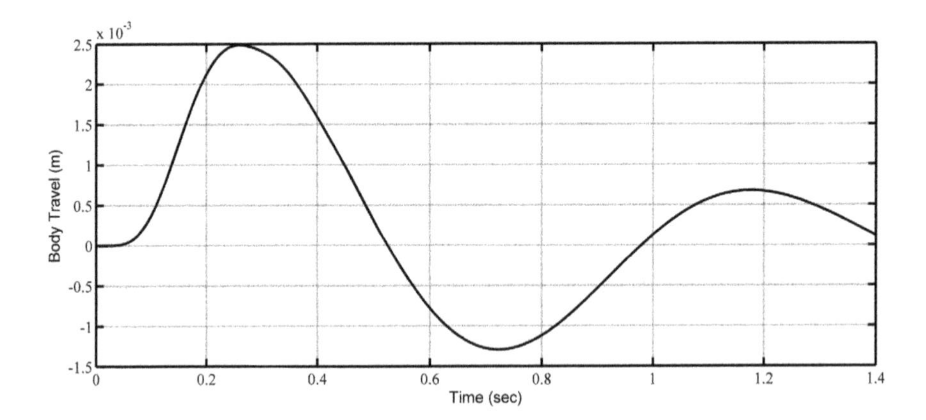

Figure 5.15 Open-loop response body deflection.

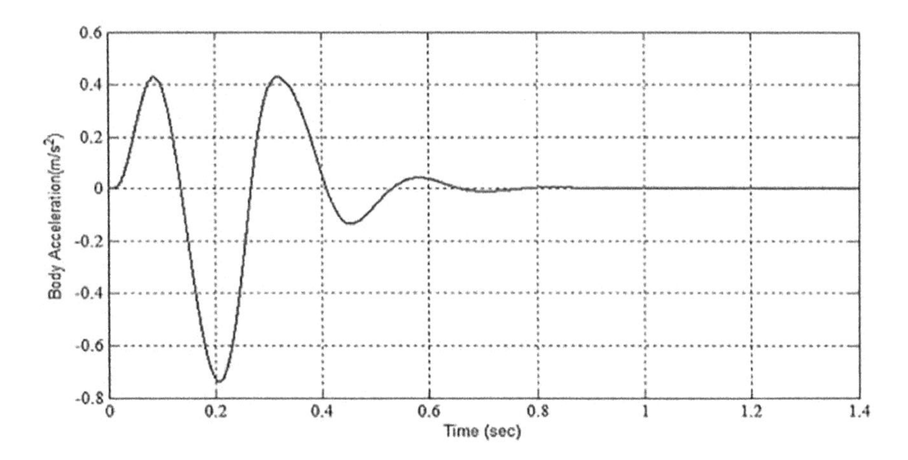

Figure 5.16 Body acceleration.

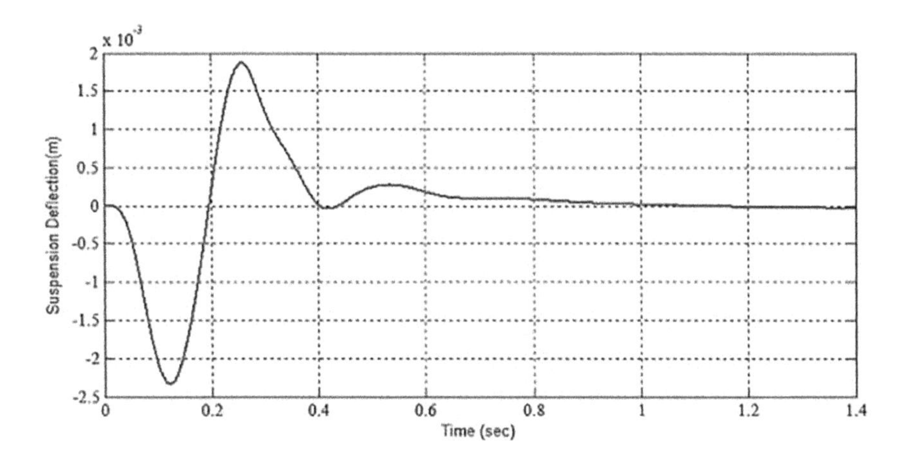

Figure 5.17 Suspension deflection.

of Body Deflection, the values of settling time and peak overshoot are found to be 1 sec and 0.00284 m, respectively. The output response of the controllers is shown in Figure 5.19.

The stability of the system can be studied using RS, NP, and RP is shown in Figure 5.20. Since the greatest value of μ is 0.46615, Figure 5.20a clearly demonstrates the RS matching the criterion for stability. Because the greatest value is 2.6026, Figure 5.20c clearly demonstrates that the closed-loop system does not attain robust performance. Thus, it can be inferred that while the developed $H\infty$ controller results in strong disturbance rejection response and system stability, it does not

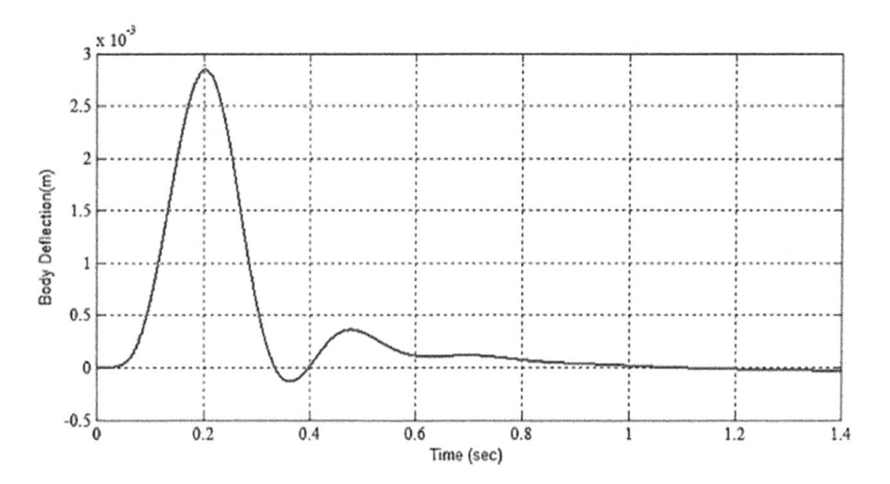

Figure 5.18 Body deflection.

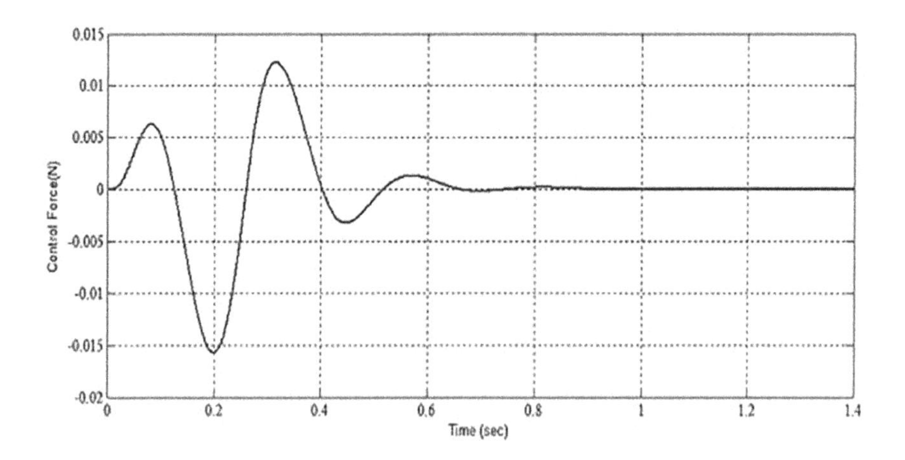

Figure 5.19 H∞ controller output.

guarantee good robust performance. There is hence a requirement for a different controller that enhances the disturbance rejection response and guarantees RS and performance.

c. *Response with μ-synthesis controller*
μ-synthesis is designed for the Quarter Vehicle Active Suspension System. The disturbance rejection responses are shown in Figures 5.21–5.23.

So, μ-synthesis controller is successfully implemented with a Quarter Vehicle Active Suspension System. In the response of Body Acceleration,

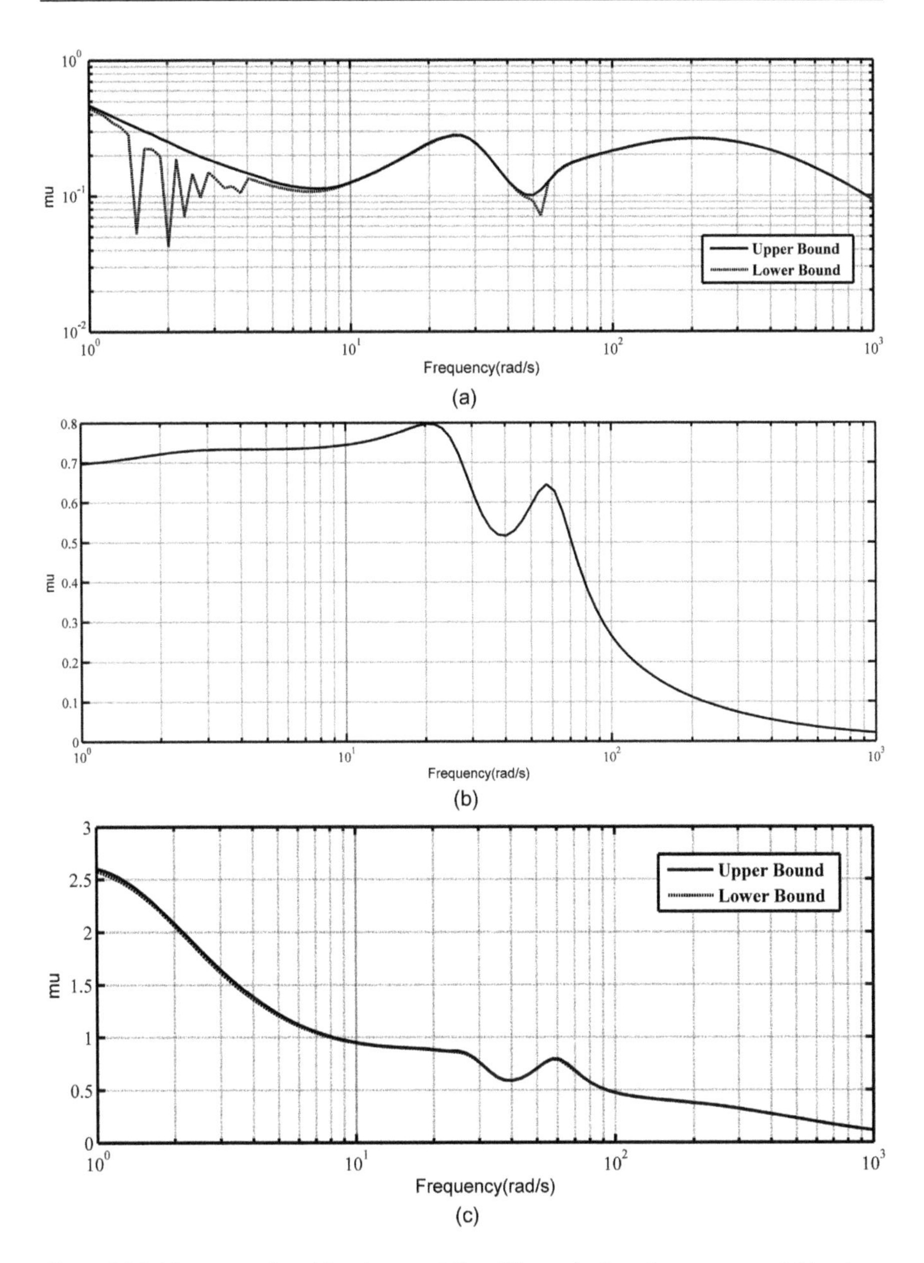

Figure 5.20 H∞ controller: (a) robust stability, (b) nominal performance, and (c) robust performance.

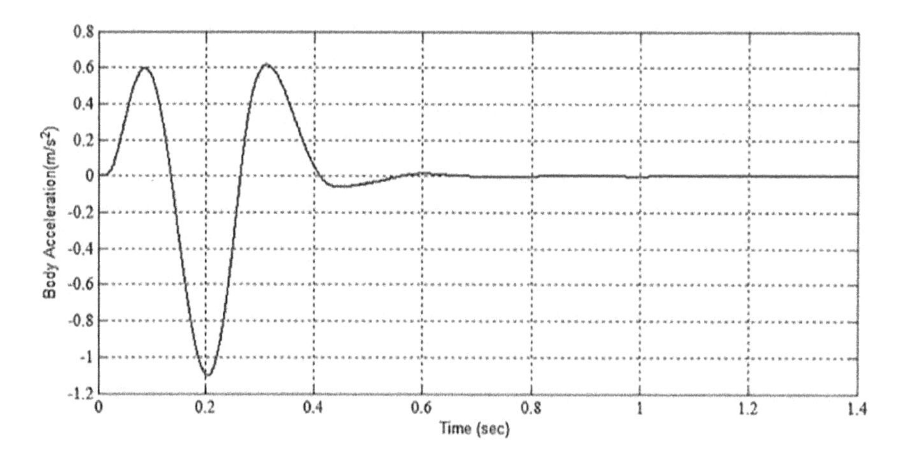

Figure 5.21 Body acceleration.

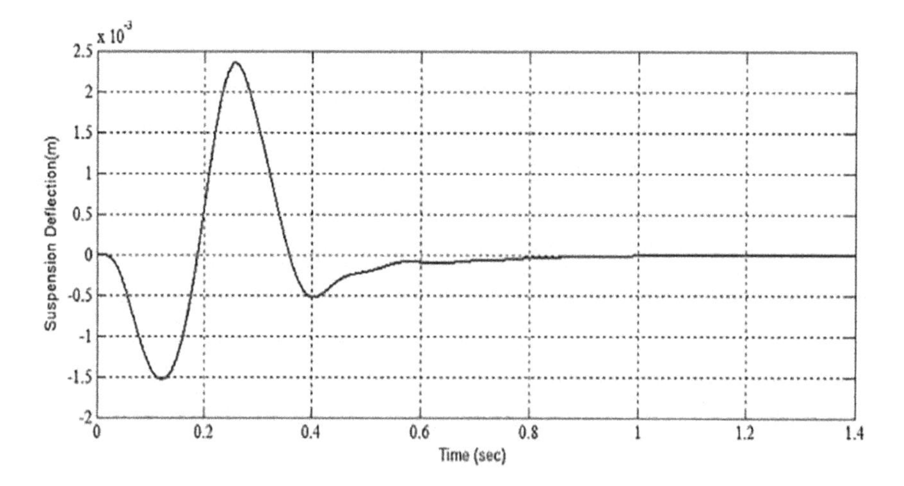

Figure 5.22 Suspension deflection.

the values of settling time and peak overshoot are found to be 0.7 sec and 0.6 m/s^2, respectively. In the response of Suspension Deflection, the values of settling time and peak overshoot are found to be 1 sec and 0.00235 m, respectively. In the response of Body Deflection, the values of settling time and peak overshoot are found to be 0.95 sec and 0.004 m, respectively. The output response of the controllers is shown in Figure 5.24.

The stability of the system can be studied using RS, NP, and RP is shown in Figure 5.25. The RS and RP of the μ-synthesis controller are depicted in Figure 5.25a and c, respectively. RS meets the criterion for stability with a maximum value of equal to 0.1585. The highest value of μ, which is

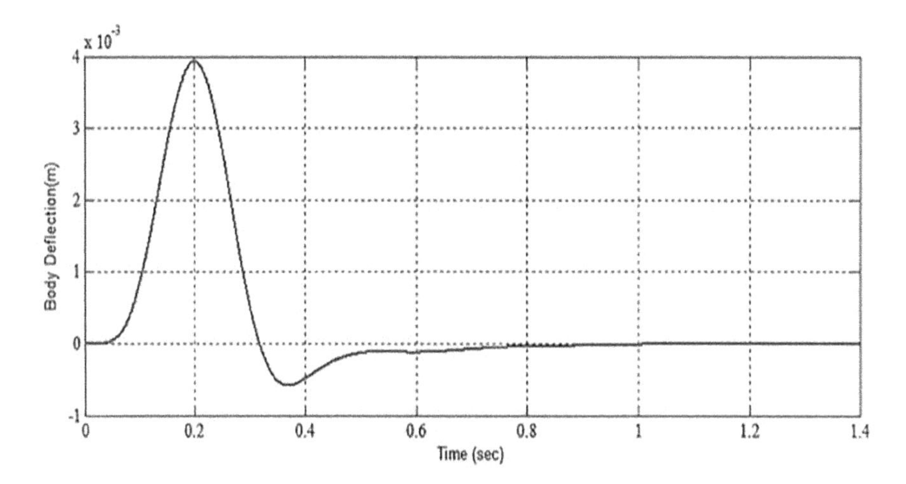

Figure 5.23 Body deflection.

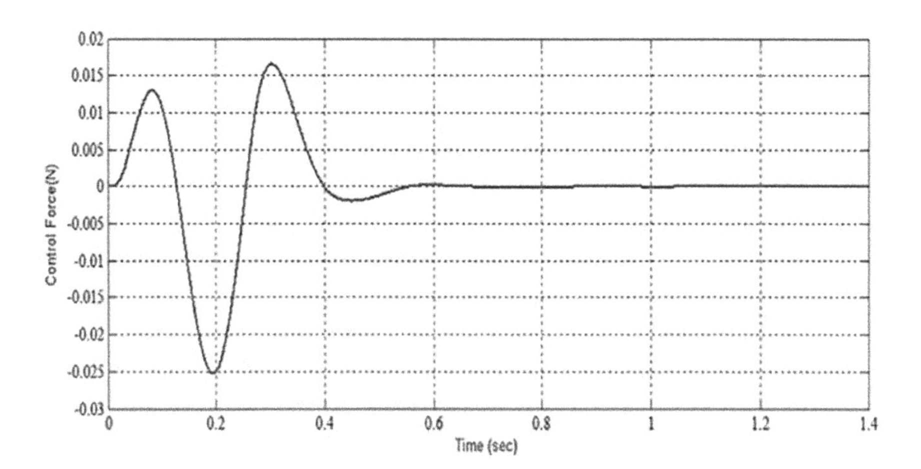

Figure 5.24 μ-synthesis controller output.

0.9089 and is less than 1, is shown in Figure 5.25c, ensuring good, reliable performance.

The time domain characteristics values of Body Acceleration, Suspension Deflection, and Body Deflection are summarized in Tables 5.2, 5.3 and 5.4, respectively.

It can be seen from Tables 5.2, 5.3, and 5.4, μ-synthesis controller is fastest having a settling time of 0.7, 1, and 0.95 sec for Body Acceleration, Suspension Deflection, and Body Deflection, respectively. But μ-synthesis controller has a larger peak overshoot as compared to the H_∞ controller for Body Acceleration, Suspension Deflection, and Body Deflection. And also steady-state error for each controller is zero.

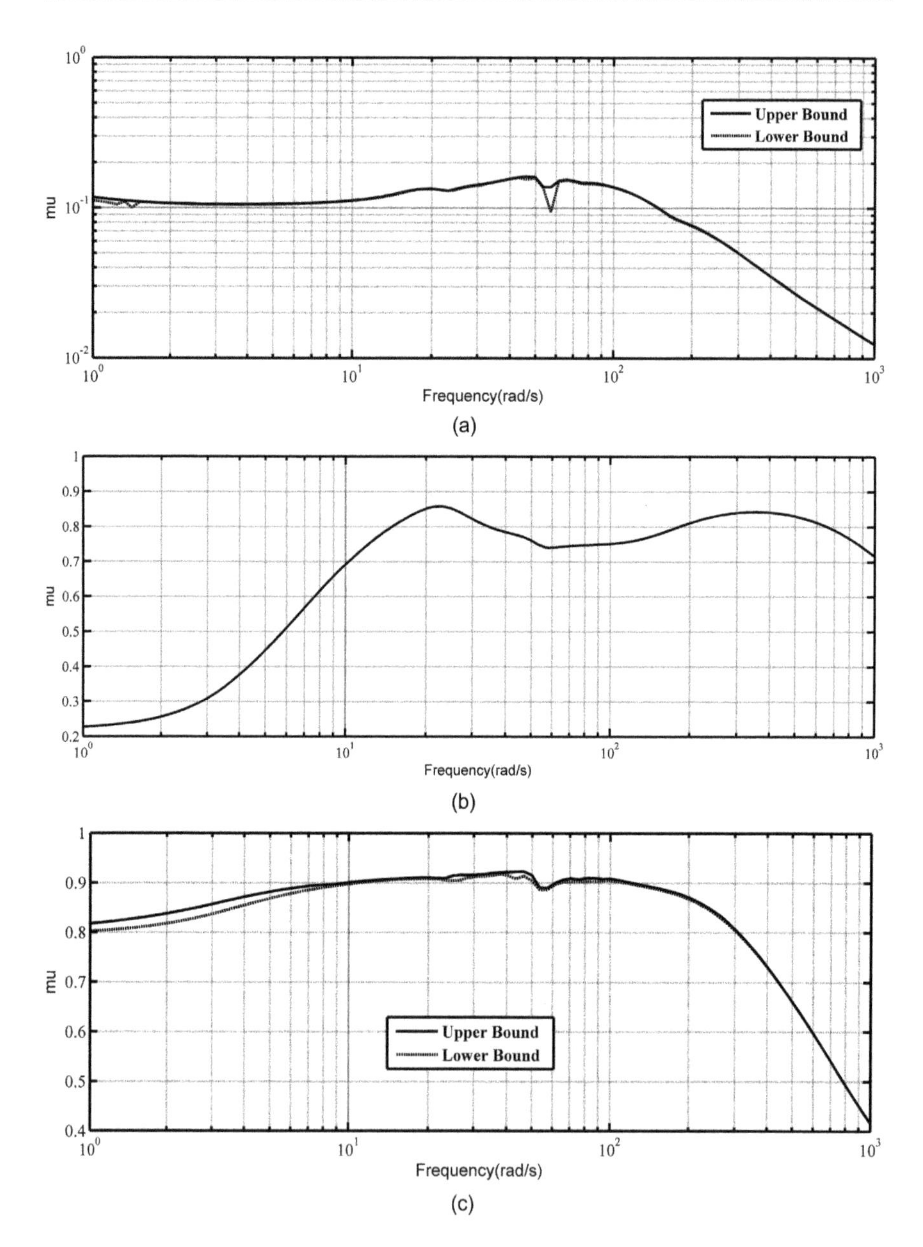

Figure 5.25 μ-synthesis controller: (a) robust stability, (b) nominal performance, and (c) robust performance.

Table 5.2 Comparison of body acceleration response

Time domain specification	*H_∞ controller*	*μ-synthesis controller*
Settling time (sec)	0.91	0.7
Overshoot (m/s^2)	0.428	0.6
Steady-state error	0	0

Table 5.3 Comparison of suspension deflection response

Time domain specification	*H_∞ controller*	*μ-synthesis controller*
Settling time (sec)	1.1	1
Overshoot (m)	0.00186	0.00235
Steady-state error	0	0

Table 5.4 Comparison of body deflection response

Time domain specification	*H_∞ controller*	*μ-synthesis controller*
Settling time (sec)	1	0.95
Overshoot (m)	0.00284	0.004
Steady-state error	0	0

5.6 CONCLUSION

To achieve comfort ride, the soft suspension system should be needed whereas stiff suspension is required to improve the road handling capability. These two conflicting objectives need to be possessed simultaneously in order to achieve desired vehicle performance. Thus, the designing and controlling of suspension system is quite challenging. The proper designing mechanism and selection of appropriate control algorithm is quite essential and also challenging due to the adversity of road conditions and other criteria. In this chapter, the quarter car active suspension model has been discussed in detail. Furthermore, the robust control strategies for an active suspension system have been presented. In this study, H_∞ and μ-synthesis controllers are successfully designed and implemented for control of a Quarter Vehicle Active Suspension System. It is concluded that the best response is obtained by a μ-synthesis control technique. A settling time of Body Acceleration, Suspension Deflection, and Body Deflection was found to be 0.7, 1, and 0.95 sec, respectively, by the use of μ-synthesis controller, which is faster than the H_∞ controller. H_∞ controller shows good RS with a slightly longer settling time while H_∞ controller does not ensure good robust performance. Finally, it can be said that both controllers are very effective at stabilizing the suspension system, but the response in the case of the H_∞ controller has a marginally longer settling time, which is enhanced when using the μ-synthesis controller. However, this comes at the expense of the peak overshoot of disturbance rejection, which is marginally subpar in the μ-synthesis controller. However, the graph of RP shows that the performance of the μ-synthesis controller is better than that of the H_∞ controller. Although both the proposed controllers have been designed successfully and are found to be stabilizing and controlling the nonlinear model of a Quarter Vehicle Active Suspension System effectively during simulation, the implementation of these techniques into real hardware is very important. Both H_∞ and μ-synthesis controllers show better performance as they are robust so they could provide better accuracy when implemented to hardware than conventional controllers which are not robust and does not consider parametric uncertainty.

REFERENCES

1. Nguyen, Tuan Anh. 2021. Improving the comfort of the vehicle based on using the active suspension system controlled by the double-integrated controller. *Shock and Vibration* 2021: 2021: 1–11.
2. Yatak, Meral Özarslan, and Fatih Şahin. 2021. Ride comfort-road holding trade-off improvement of full vehicle active suspension system by interval type-2 fuzzy control. *Engineering Science and Technology, an International Journal* 24 (1):259–270.
3. Hać, A. 1985. Suspension optimization of a 2-DOF vehicle model using a stochastic optimal control technique. *Journal of Sound and Vibration* 100 (3):343–357.
4. Kumpati, S. Narendra, and Parthasarathy Kannan. 1990. Identification and control of dynamical systems using neural networks. *IEEE Transactions on Neural Networks* 1 (1):4–27.
5. Mahmoodabadi, M.J., and N. Nejadkourki. 2022. Optimal fuzzy adaptive robust PID control for an active suspension system. *Australian Journal of Mechanical Engineering* 20 (3):681–691.
6. Ovalle, Luis, Héctor Ríos, and Hafiz Ahmed. 2022. Robust control for an active suspension system via continuous sliding-mode controllers. *Engineering Science and Technology, an International Journal* 28:101026.
7. Hrovat, D., and M. Hubbard. 1987. A comparison between jerk optimal and acceleration optimal vibration isolation. *Journal of Sound and Vibration* 112 (2):201–210.
8. Khargonekar, Pramod P. 1987. Control system synthesis: A factorization approach (M. Vidyasagar). *SIAM Review* 29 (4):658–660.
9. Sharp, R.S., and D.A. Crolla. 1987. Road vehicle suspension system design: A review. *Vehicle System Dynamics* 16 (3):167–192.
10. Smith, Malcom C., and Gavin W. Walker. 2000. Performance limitations and constraints for active and passive suspensions: A mechanical multi-port approach. *Vehicle System Dynamics* 33 (3):137–168.
11. Karnopp, D. 1995. Active and semi-active vibration isolation. *Journal of Vibration and Acoustics* 117 (B):177–185.

12. Hrovat, D., D.t. Margolis, and M. Hubbard. 1988. An approach toward the optimal semi-active suspension. *Journal of Dynamic System, Measurement, and Control* 110 (3):288–296.
13. Akçay, Hüseyin, and Semiha Türkay. 2011. Influence of tire damping on actively controlled quarter-car suspensions. *Journal of Vibration and Acoustics* 133 (5):054501.
14. Chalasani, Rao M. 1986. Ride performance potential of active suspension system-Part 1. *ASME Monograph* 80 (2): 187–204.
15. Fialho, Ian, and Gary J Balas. 2002. Road adaptive active suspension design using linear parameter-varying gain-scheduling. *IEEE Transactions on Control Systems Technology* 10 (1):43–54.
16. Jin, Xianjian, Jiadong Wang, Shaoze Sun, Shaohua Li, Junpeng Yang, and Zeyuan Yan. 2021. Design of constrained robust controller for active suspension of in-wheel-drive electric vehicles. *Mathematics* 9 (3):249.
17. Singh, Narinder, Himanshu Chhabra, and Karansher Bhangal. 2016. Robust control of vehicle active suspension system. *International Journal of Control and Automation* 9 (4):149–160.

Chapter 6

Electric vehicle energy management

An intelligent approach

Bakul Kandpal and Ashu Verma
IIT Delhi

CONTENTS

6.1 INTRODUCTION

Decarbonization of the energy sector can put additional stress on traditional electric-power system operation. In particular, the envisioned modern transportation system under electric vehicles (EVs) would require expansion of generation power and transmission and distribution networks [1]. The demand of EVs follows peculiar spatio-temporal charging patterns. In particular, historical data for EV charging shows that a surge in charging demands occurs at either evening or morning hours [2]. This impact of EV charging would primarily be seen at the distribution level, specifically in the residential and commercial settlement areas. This would in turn imply that a low-voltage network would be susceptible to a surge in peak demands and thereby under-voltages and high power losses. These impacts would also be impeding the operation, control, and stability of transmission systems.

However, energy management systems (EMS) for EVs can give time-based or location-based coordination of charging EVs. This would induce demand-side management or demand response (DR) of EV power consumption, allowing throttling of EV power consumption to adjust quick charging requests, or to increase the utilization of renewable power consumption. DR strategies, aiming at demand-side management, can control aggregated power demand of EVs in synchronizing to the grid issues, such

DOI: 10.1201/9781003436089-6

as frequency or voltage problems. The issues arising out of EV penetration at the distribution level at smart control methods to mitigate them are discussed in the next sections.

This chapter is organized as follows: First, in the Introduction section, the adverse impact of uncoordinated EV charging demand on the power sector is discussed. Moreover, the idea of demand-side management and communication and centralized database requirements for EV scheduling are also presented. Next, the DR strategies under a variety of ancillary objectives are briefly addressed, while the state-of-the-art uncertainty handling techniques mitigating an adverse impact of renewable generation on EV charging behaviour is also discussed. Finally, modelling of decentralized EV energy management strategies is presented, which aims at improving user privacy and algorithmic scalability of DR under a large number of EVs.

In several countries of the world, sales of EVs have been reaching landmark acceptance, while the purchase of diesel vehicles is gradually decreasing [3]. At the end of 2021, there were around 16.5 million battery-operated EVs on the road [4]. For instance, the sales of EVs have seen a steep increase in China and the US (especially California). In Norway, due to tax incentives (CO_2 emission component), more than half of all new vehicles sold are electric. To achieve the ambitious global targets on net-zero carbon emissions, countries around the world are prohibiting sales of non-EVs while allowing incentivization for charging infrastructure deployment.

Along with alternative fuels such as hydrogen, it is established that EVs would be a major share of the consumer transportation industry [5]. The use of EVs would dominate the light and heavy duty segments, including that of electric buses/E-buses, and long-haul logistics services provide by electric trucks [6]. Moreover, in Scandinavian countries, marine transportation is also planned to be transitioned to hybrid or electrified solutions to reduce dependency on marine gasoil [7].

With a wide variety of EVs requiring power from the grid, the infrastructure development will also be varied. That is, the EV supply equipments (EVSEs) can be single-phase ranging from lower than 3.2–7.4 kW power output [8]. In contrast, EV charging through three-phase supply equipments can go typically as high as 15 kW per EVSEs, while ultra-fast DC chargers can be of capacity up to 350 kW [9]. Moreover, heavier vehicles such as E-buses require large battery packs to attain sufficient travel distance to complete their driving tasks. Therefore, they are equipped with larger battery sizes, typically of 200 kWh and require charging infrastructure with an average 100 kW power supply [10]. For more convenient delivery of power, wireless charging methods at bus depots/stops are also studied in pilot projects around the world, which aim at reduced battery size requirement of E-buses [11].

Integrating charging facilities, especially high power chargers for fast charging, would alter the power flow in the network. Moreover, bidirectional power flow from discharging batteries can cause over-voltages in the network. In a conventional distribution network, the controllable resources are few, which only allows the corrective measures to be adopted at the transmission side. Moreover, charging of EVs can create additional burden on the network with intermittency and unpredictable surge in power consumption creating recurring peaks. It is well established that EV charging demands are unpredictable and in conjunction with intermittent photovoltaics (PV) generation can cause forecasting errors and may force network operators to perform voltage corrections through additional infrastructure upgrade [12]. Such demand perturbations can also cause stability issues in the grid [13]. To ensure voltage stability, the voltage at all buses of the network should be able to tolerate a disruption under standard operating conditions [14]. The challenge of voltage stability issues can further deteriorate due to ill-formed DR strategies for EVs or distributed energy resource (DER) generators [15]. The transition to a sustainable energy future with EVs and DERs would also result in complex load and generator dynamics, which would cause low redundancy and error of measurements of network's system variables [16]. Moreover, uncoordinated consumption of EVs can cause congestion of lines and higher losses [17].

The power flow in a radial distribution network can be modelled as follows: Denote the power injections at each node $j \in V$ as $P_j + i\,Q_j$, the resistance and reactance of each power line $l \in E$ can be denoted as $Z_l = R_l + i\,X_l$. The power flow equations can be modelled in a linear form using *LinDistFlow* equations as follows [18]:

$$P_{\{j+1\}} = P_j - p_{\{j+1\}} \tag{6.1}$$

$$Q_{\{j+1\}} = Q_j - q_{\{j+1\}} \tag{6.2}$$

$$V^2_{\{j+1\}} = V^2_{\{j\}} - 2(re_j\ P_j + \{xe_j\}Q_j \tag{6.3}$$

where the voltage of a bus j is modelled using (6.3).

Therefore, the charging consumption of EVs should be limited such that a constraint on power limit of each line would not be violated. Thus, location and time of charging can be key in impeding a negative impact of EV adoption on power grids. This is especially true for networks that are constrained on transmission and distribution infrastructure. Moreover, the losses on every line would increase with increasing power flow. Therefore, the demand and generation supply mismatches can occur in the grid due to uncompensated losses. The increased power losses would also imply a non-linear increase in generation costs as modelled through generation cost curve. The increase in generation cost would thereby result in increased charging costs to EV users. The concept of locational marginal pricing or shadow prices is commonly used in the literature to define the impact of network constraints on electricity tariffs at any node/bus [19]. In this regard, loss sensitivity factors in the network can ascertain the economic impact of power losses or increase in locational marginal pricing due to losses as follows:

$$\frac{\{\partial P_{\{loss\}}\}}{\{\partial P_k\}} = 2\,Re\left[\,G\{\partial f\{V\}\}\{\partial P_k\}\right] \tag{6.4}$$

which defines the loss in the network due to an increase in power demand (denoted as P_k) at any bus/node k. It may be noted that the sudden increase in power consumption of EVs would also be detrimental to the financial objective of EV owners. This is because the distribution system operator (DSO) can therefore increase the tariff for any charging station, thereby transferring the increased cost of generation to the EV owners. This phenomenon has been studied in the previous literature under sensitivity analysis of optimal dispatch of generators [20].

Moreover, single-phase EV charging can cause uneven power in flow in the three phases. As power losses are non-linearly related to the current flow in the line, phase load unbalancing (PLU) can result in increased power losses as compared to a balanced grid. The neutral wire would be heavily loaded under PLU and transformer and line thermal limits can be exceeded, requiring expensive infrastructure upgrade by the DSO to cater to uncoordinated single-phase EV charging.

Formally, the PLU among the three phases can be formulated as

$$\max_{\{i\in\Omega\}}\left(\{\psi_i - \psi_{\{avg\}}\}/\{\psi_{\{avg\}}\}\right) \tag{6.5}$$

where $\Omega = \{R, Y, B\}$ defines the set of three phases. In (6.5), the maximum deviation of any phase's load from the average $\psi_{\{avg\}}$ is computed.

As discussed in the previous section, the modern transportation system including EVs would require planning for their inclusion in conventional transmission and distribution networks. For additional study, a more detailed review of adverse impact of EV to grid integration is provided in [21]. However, with the coordination of EV charging demands, many of the problems such as recurring peaks or under-voltages can be avoided. This would require infrastructure upgrade for EV charging, especially for control signals between DSO, EV aggregators (EVAs), EV owners, and EVSEs. The signal communication can be performed under open-source protocols such as open charge point protocol (OCPP) [22]. Such protocols are enforced in some countries for all charging communication requirements between charging stations and a central authority for data management and storage [23]. The communication between the EVs and the server would require a specific protocol to share information. ElaadNL published an EV-Related Protocol Study in [24], which presents a discussion of a selection of EV-related communication protocols. The study gives insight as to how the protocols could be used and combined in the future for communication. The study states that in future, the state of charge (SOC) (and perhaps the departure time as well) can be acquired from an EV using the ISO/IEC 15118 communication protocol. Moreover, ISO 15118 also states the standards for vehicle to grid (V2G) communication interface between vehicle and an EVSE (Figure 6.1).

Therefore, the desired energy (in kWh) from the PEV to the EVSE can be sent using OCPP 2.0 protocol, and the EVSE can then forward this information to the back-end server. Thus, such communication systems can be used to link the EVs to the servers for information sharing. The use of communication protocols unifies vehicles, charging stations, communications, and networking systems work in unison with the electric grid. Therefore, the transportation sector should expand the use of open and inter-operable communication standards and open-source software for lowering the cost of public charging solutions.

On the other hand, the charge acceptance of EVs would also depend upon the charging protocols used by the battery management system of the EV. BMS charging protocols are used to manage the electrical current sent to the EV batteries based on SOC or temperature. This helps in saving the batteries from over or under charging and cyclic or temperature-based degradation. The most common charging protocols are constant current (CC), constant voltage (CV), CC-CV, and multistage constant current (MSCC) charging protocols [25]. As the name suggests, CC protocol allows only a small constant current throughout the charging process, while CV protocol allows only charging at a constant voltage. The combination of both CC-CV allows only a constant current till a pre-defined SOC of battery (of around 80%) then gradually drops the current sent to the battery. In contrast, the MSCC protocol has a fixed current charge limit allowed for a particular range of battery SOC.

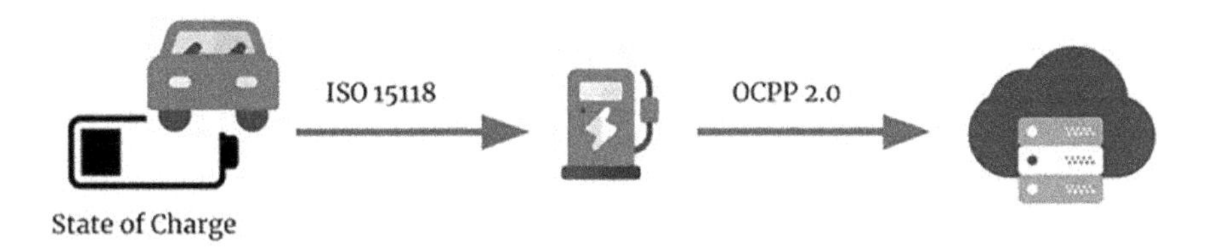

Figure 6.1 EV feedback signal implementation in real world. Information on EV battery SOC is communicated through EVSE to the cloud database using ISO 15118 and OCPP protocol.

Finally, the EVSE can help in four-quadrant control power flow, thereby having the potential to provide ancillary service to the grid [26]. In particular, the reactive power injection can help support voltages in the network. This is especially true when the inverter capability of an EV charger is not being fully utilized. It is known that inverters for EV chargers would have a capability curve as

$$P = \sqrt{\{S^2 - Q^2\}} \tag{6.6}$$

where S is the inverter capacity in kVA and P and Q define the active and reactive power consumption of the EV. It may be noted that bidirectional chargers would allow four-quadrant control through injection of negative P and Q. Such throttling of power flow can help in providing ancillary services such as local reactive power consumption, voltage deviation minimization, and loss reduction [27].

6.2 SMART CHARGING STRATEGIES FOR EVs

With appropriate incentives, the energy storage capability of EV batteries can be utilized to reduce the adverse impact of EV-grid integration. In addition, an aggregator (or EVA) can use smart charging strategies for managing the EV battery pool to provide ancillary services for power distribution systems. In effect, DR programs can be initiated that allow time-based charging schedules of EVs to reduce charging peaks, etc. Further, the use of aggregated EVs can be performed as a storage system to supply power in time of peak demand or disruption in generation. Therefore, EV batteries can be discharged based on pre-decided dispatch signals. The framework for implementing a DR program focused on EV scheduling is shown in Figure 6.2.

In the same context, the rapid and large-scale deployment of various renewable energy systems have considerably impacted the structure of grid by moving from conventional generation to variable distributed energy production. The challenge of developing smart EV charging strategies lies in understanding the need of EV owners and grid operators, combining it with prediction on EV consumer and grid system variable uncertainties. Considering the charging variable of an EV 'n' as p_n, an objective of DR can be formulated as

$$\min f\left(\{\boldsymbol{P}\}, p_n\right) \tag{6.7}$$

where the function 'f' is dependent upon the charging variable p_n and a concatenated variable **P**, denoting the variables of all other EVs. The function may denote a range of objectives such as charging cost reduction on

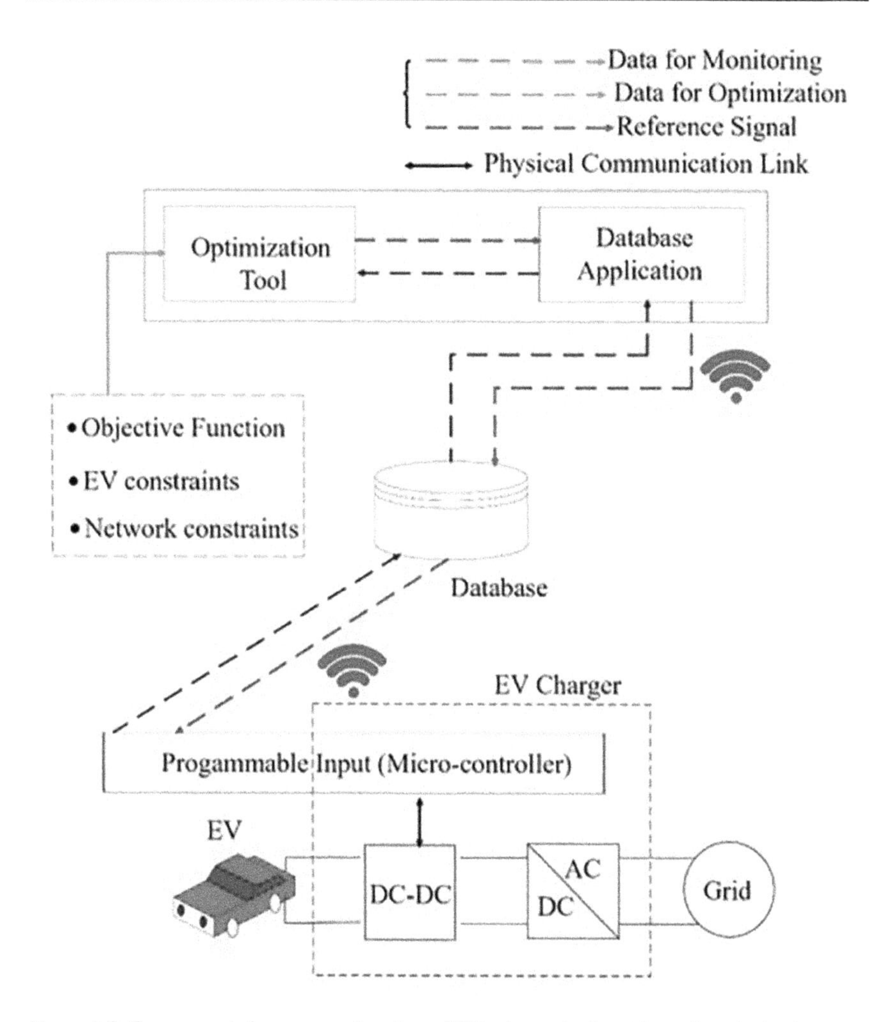

Figure 6.2 Framework for smart charging of EVs through cloud-based optimization.

the basis of time-of-use prices, day-ahead prices, or real-time prices. The electricity prices can be obtained from day-ahead market clearing where hourly prices would differ in peak and off-peak hours. The charging of EVs can then be achieved with an objective as follows [28]:

$$f(\{P\}) = \{c\}^T f\{\boldsymbol{P}\} \tag{6.8}$$

where **P** denotes the vector of charging variables for all EVs, while the hourly price is denoted by the vector $\boldsymbol{c}$.

The challenge with price-based charging can be of recurring peaks, where cheaper hours would invite high charging demands and thereby become new peaks. To cater to such problem, the charging of EVs (or battery storage systems) can be performed specifically based on peak and off-peak hours, as in [29],

$$f(\{P\}) = |\{P\} - L_{\{avg\}}|_2^2 \tag{6.9}$$

where $L_{\{avg\}}$ denotes the average load of the entire day. This would ensure that EV charging is mostly in valley periods, i.e., periods with low demands. However, the shift of EV charging can also result in degradation of EV batteries due to sudden increase or decrease in their SOC. The sudden change in SOC can be controlled such that the degradation of batteries is limited as follows:

$$f(\{P\}) = \sum P_i(t)\ \Delta t \tag{6.10}$$

where t is the time index and r denotes the non-linear battery capacity fading rate with each cycle [30].

Moreover, the degradation of batteries would be majorly seen if discharging is also allowed. That is, the variable $\boldsymbol{P}$ would be allowed to be negative,

$$P^{\{min\}} \prec \{\boldsymbol{P}\} \prec P^{\{max\}} \tag{6.11}$$

where $P^{\{min\}}$ is a negative constant and '$\prec$' denotes the generalized inequality over $\{\boldsymbol{R}\}^n$. The discharging of EVs can be to the grid, i.e., in V2G mode, or directly supply power to a building or a house in vehicle to building (V2B) and vehicle to home (V2H).

While managing energy consumption of EVs, it should also be noted that individual EV owners would have their specific SOC requirements and maximum time delay. The SOC requirements of individual EVs can be modelled as

$$SoC_{\{n,\ t+1\}} = SoC_{\{n,\ t\}} + \eta_c\ \left(\{p_{\{n,\ t\}}^{\{EV\}}\ \Delta\ t\}\right) / \{B_n^{\{ev\}}\} \tag{6.12}$$

$$\{SoC_{\{n,\ t\}}^{\{min\}}\} \le SoC_{\{n,\ t\}} \le \{SoC_{\{n,\backslash,\ t\}}\} \tag{6.13}$$

$$SoC_n^{\{\ req\}} \le SoC_{\{n,\ T\}} \tag{6.14}$$

$$0 \le p_{\{\ t,n\}}^{\{EV\}} \le \{p_{\{\ t,n\}}^{\{EV\}}\} \tag{6.15}$$

where (6.12) denotes the SOC update with time, (6.13) ensures SOC to be within limits, and (6.14) fulfils the individual energy requirement of each EV till its departure time T.

6.3 UNCERTAINTY HANDLING IN DR ALGORITHMS

One of the major challenges for EV management strategies is of unforeseen charging developments unaccounted in DR. The charging consumption pattern of EVs is unpredictable because their arrival and departure at a parking lot can be sporadic. Some EV owners would suddenly require to disconnect their EV, in-between of the pre-planned charging duration, to depart. Such actions can remove some EVs from the parking lot without prior notice of the EVA. Therefore, it is imperative that EV charging demand is modelled as a random variable. In particular, the arrival or departure of EVs at a parking lot can be modelled as a Poisson distribution [31],

$$Pr(X = x) = e^{\{-\lambda\}}\{\lambda^x\}/\{x!\} \tag{6.16}$$

where '*k*' is the number of EV arrivals/departures. Thus, the probability of arrival for a fixed number of EVs x can be found. Therefore, an EVA can generate several scenarios of EV arrival, each with a certain probability. Using this information, the EVA can solve an expected objective with a wide variety of pre-found scenarios. As the generated scenarios can be large, for further analysis scenario-reduction techniques such as *k-means* can be employed to remove any redundant information in the scenario modelling process. After scenario modelling, the EVA can formulate a stochastic objective as follows:

$$min\ \boldsymbol{E}\ f(\{P\}) \tag{6.17}$$

where the expectation is taken over all the scenarios. It can be noticed that using Jensen's inequality, we have

$$f(\boldsymbol{E}\{P\}) \le \boldsymbol{E}[f(\{P\})] \tag{6.18}$$

which means that solving the deterministic EV scheduling objective would give the lower bound of the stochastic problem in (6.17), and the expected value of objective under uncertainty would be greater than its deterministic counterpart. Moreover, a common extension of stochastic problems used in scheduling algorithms is of two-stage planning. That is, the scheduling or energy management is divided into two stages and different set of variables are used which are bound by a constraint. The mathematical formulation of a two-stage stochastic objective can be written as

$$\min\ \boldsymbol{c}^{T}\ P + \boldsymbol{E}\ Q(P, P') \tag{6.19}$$

$$Q(P, P') := \min q^{T}\ P' \tag{6.20}$$

$$h(P, P') = 0 \tag{6.21}$$

where P and P' are the variables for the first and second stages, respectively. Moreover, in (6.19), the two variables are bound by a common constraint. Such two-stage stochastic problems can arise if the EV scheduling problem can be divided into parts. That is, it can be assumed that the first set of variables $\boldsymbol{P}$ are used for day-ahead planning, whereas P' are used for real-time scheduling planning of EVs. If an EVA follows such two-stage planning approach, the uncertainty in forecasted data, as in day-ahead markets, can be mitigated.

The use of DERs such as PV would also be beneficial in mitigating the adverse impact of fast-charging EVs. However, PV generation output is itself uncertain due to random solar irradiation throughout the day ahead. The uncertain solar radiation for any future time can be formulated as a Beta distribution,

$$f(R_t) = \Gamma(\alpha_t + \beta_t))\} / \{\Gamma(\alpha\_t\ )\Gamma(\beta\_t)\}\hat{}\{(\alpha\_t - 1)\}(1 - R_t)\hat{}\{\beta_t\} \tag{6.22}$$

where the parameters can be found using historical data [32]. Therefore, a similar scenario generation strategy can work for mitigating uncertainties in PV generation also. However, a major drawback of stochastic modelling lies in inaccurate probability density functions for denoting uncertainties.

The stochastic strategies of DR are used to minimize the expected solution given the variety of scenarios. However, a challenge in such an approach could be that the parameter variance in scenario modelling for optimization would be too high. Therefore, a simple stochastic scheduling algorithm would consider extreme scenario values, which can result in poor capture of uncertainty. Moreover, if the objective value function is normally distributed around the mean, the solution of deterministic and stochastic would be nearly equal. A workaround to this would be the use of confidence intervals for uncertainty. The use of confidence intervals would allow minimization of conditional value at risk (CVaR). The CVaR gives the minimum value of the objective function under uncertain input parameter, which satisfies a certain probability $\beta \in (0,1)$ for the uncertain parameter in the optimization model. In our EV scheduling algorithm, value at risk would signify the minimum value of objective function at a given probability of

PV generation at β or above. In summary, the CVaR would minimize the conditional expectation of the EV scheduling objective given the confidence level β for uncertain parameters such as PV generation,

$$\min \phi_{\beta}(P) \tag{6.23}$$

$$\phi_{\{\beta\}}(P) = (1-\beta)^{\{-1\}} \int_{\{f\geq\alpha\}} f\left(P, y^{PV}\right) dy^{\{PV\}} \tag{6.24}$$

where the parameter for PV generation output $y^{\{PV\}}$ is random with probability density function $p\left(y^{\{PV\}}\right)$. For additional details on optimizing objective functions using CVaR, the reader is referred to [33].

In contrast, a different approach to modelling uncertainties is of worst-case minimization. For instance, the worst-case charging cost based on prices can occur if cheaper tariffs in day-ahead markets are much more expensive in real-time settlement. This can occur due to a sudden generation outage or contingency in the network, etc. A robust energy management strategy for EVs would thereby signify resilient operation even under the worst-case possible future. That is, the EV energy management objective would be transformed to

$$\min \max_{\{u\in \upsilon\}} \quad f(P,u) \tag{6.25}$$

where u is the uncertain input parameter, such as the hourly prices, whereas υ is the uncertainty set. The uncertainty set has to be pre-found by the DR operator such that it captures the correlation between uncertain parameters. As an example, in price uncertainty, the uncertainty set υ would be the maximum and minimum limit of hourly prices, i.e., $c_{\{\max\}}$ and $c_{\{\min\}}$. It should be noted that robust optimization gives highly conservative results, especially if the uncertainty sets are not properly designed [34].

Similar to the previously mentioned two-stage stochastic approach, an adaptive robust strategy can be modelled as

$$\min \ c^{T}P + \max_{\{u\in\upsilon\}} f(P',u) \tag{6.26}$$

$$h(P, P') = 0 \tag{6.27}$$

where we divide the concatenated variables into two stages P and P'. This would give the EVA the freedom to choose second-stage variable after additional information on the uncertain parameter is unfolded.

Management of a large fleet of EVs would only be possible if the exchange between the EV agents, EVA, and the DSO is economically beneficial for all. In particular, a DR program would only be economically feasible if it satisfies the following simple issues:

- Every agent should not have financially better alternatives for participation elsewhere.
- The procurer of energy (such as the DSO or EVA) should at least be able to achieve levelized costs for energy supplied.
- Higher participation, such as by EV owners, should be rewarded by higher incentives.

To ensure the aforementioned points, the interaction between EVAs, EVs, and DSO can be formulated with game theoretic approaches. As decisions of all agents are interdependent, the final output can only be made by recurring negotiations between players. In particular, non-cooperative interactions are settled in iterations, where each iteration would denote a DR bid by an individual participant.

For instance, the EVA may take the lead and decide a marginal price for charging of all EVs under its purview. After price broadcast, the EV owners at the station would respond by changing their demand. This assumes that the charging demand of EV owners is elastic, varying as per the prices. Moreover, the procurement cost of energy for the EVA would directly be dependent upon the charging demand of EVs. Therefore, the EVA would revise its prices based on the new demand requirement bids by the EVs. This sequential bidding between EVs and EVA would continue till all players or agents do not change their bids [35]. Non-cooperative interactions are often shown to converge to an equilibrium, when sequential bids by all participants conclude. It is straightforward to reason, by definition, that no participant can do any better from moving away from the bids achieved at equilibrium.

The EVA can also convince the EVs for jointly using their batteries in a coalition, as shown in Figure 6.3. Under coalition games, a common objective is decided by all agents. All EVs would thus work on a unified objective, such as voltage improvement or peak reduction. The profit incentives or costs of DR are then distributed to all EVs in a fair and rational way [36]. However, a major challenge in forming coalitions is ensuring fair incentives or cost distribution to all participants, especially taking into account their individual contributions towards the program objective. If any smaller sub-coalition of EVs can deviate away from the larger coalition to attain increased financial benefits, cooperative coalitions under centralized scheduling would not work. This could happen in cases where the incentive distribution does not ensure individual and sub-coalition rationality. Moreover, if the formulation of EV energy management objective

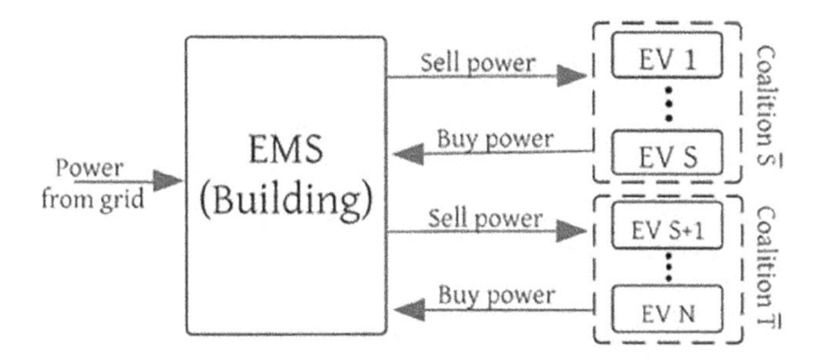

Figure 6.3 A coalitions of EVs, parked at a building, participating in DR for optimal exchange of energy.

is not strictly convex, the solution of EV scheduling algorithm would not be unique, having several optimal solutions. In such a case, some EVs may become under or over utilized with the DR program, thereby reducing their financial incentives from the optimal. The advantage of using coalitions is that computational units are reduced because of a central authority, such as an EMS in Figure 6.3, makes the charging decision for all EVs. In contrast, in non-cooperative EV energy management, every agent, i.e., every EV user would require a separate computational unit to find its own demand bids. In summary, centralized frameworks would prefer cooperative scheduling of EVs, whereas non-cooperative scheduling would require decentralized or distributed framework.

6.4 DECENTRALIZED ENERGY MANAGEMENT FOR EVs

EV scheduling for management of charging (and discharging) demand would require substantial computational requirement. In a central computational server, the burden for computing will be aggregated at one place. A decentralized computational framework can divide the optimization into several devices at once. Moreover, using decentralized scheduling, the optimization can be performed in parallel and the loss of a single computational device will not halt the operation of DR.

Ideally, the distributed DR scheduling strategy aims to find every EV's optimal charge consumption using an individual CPU. This would be feasible in cases where each EV is connected to a particular house, with several houses participating in DR. Therefore, each EV would have its individual objective function f_i and its individual set of energy constraints. In a community, the EVA may want to constrain the aggregated charging demand of all EVs, specifically to avoid violation of network constraints. In particular, the mathematical formulation of a conventional EV distributed scheduling algorithm would be as described in [37], where, EV minimizes its own Lagrangian by keeping the charging variables of all other EVs, i.e., the set of EVs $N - i$ as constant, obtained from previous feedback. The dual parameter is updated using a centralized update server as shown in Figure 6.4. The update of solutions can also be performed in the fully decentralized framework, with parallel update of solutions from all EVs [38]. Parallel update of DR solutions can be achieved using a proximal term in the Lagrangian $\left| p^{\{K+1\}} - p^{k} \right|_{S}$, which limits the deviation in subsequent solutions of the decentralized algorithm. It should be noted that convergence and optimality of decentralized scheduling algorithms depend upon properties (such as convexity, etc.) of the underlying optimization framework [37].

A supplementary advantage of decentralized scheduling is improved privacy of user power consumption data. That is, decentralized scheduling such as that through alternating direction method of multipliers (ADMM) does not send private user energy requirements to a central database for computation, as was earlier proposed earlier and in Figure 6.3. In contrast, each EV agent only shares its iteration-wise schedule to the parameter update server. However, iterative information sharing over an ADMM algorithm can give rise to privacy concerns, especially if user-sensitive data is proprietary to each consumer [39]. User privacy can become an obstacle for acceptance of DR algorithms, as user profiling can identify specific information on EV owners such as their presence at home, future trip information, or information about places of work [40].

To improve user privacy, two major types of data privatization techniques can be employed, namely, encryption based, i.e., blockchain, etc., or noise-based, i.e., differential privacy, etc.

Moreover, as elaborated in [41], blockchain-based bidirectional communication between EVs and a coordinating server will have some stringent requirements such as fast authentication, limited access of data to the certificate authority and protection against man-in-the-middle and replay attacks. In contrast, differential privacy uses noise infusion in the private data such that the information becomes randomized. Figure 6.5 shows two different methods to make the EV charging data differentially private decentralized algorithms. In Figure 6.5a, dual perturbation with variable 'y' infused with noise by coordinator is shown. However, dual perturbation

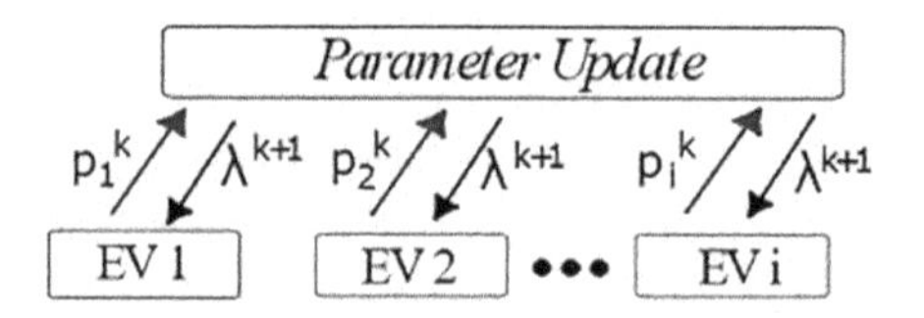

Figure 6.4 Coordinated gather and broadcast for dual update.

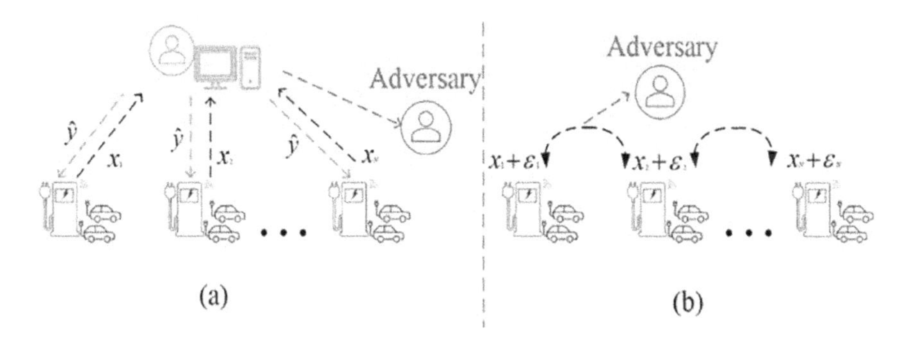

Figure 6.5 Distributed EV scheduling under differential privacy.

would still require a parameter update server and thus would not qualify for fully decentralized scheduling. In contrast, Figure 6.5b shows fully decentralized EV scheduling with individual agents appending noise ξ_i to their own schedules x_i. The noise to be added is picked from a Laplacian distribution. That is, using L_1 sensitivity, a random variable is sampled from a Laplacian distribution centred at zero and scale Δ / ϵ .

Although, the performance of ADMM will suffer from noise addition. As noise is purely from each EV agent, the convergence properties of the algorithm can become worse, especially if the variance of noise is large. The mitigation of noise-based uncertainty can be achieved by using Bayesian or uncertainty modelling techniques as discussed in previous sections.

6.5 CONCLUSIONS

The paradigm shift towards carbon neutrality would favour exponential growth of EVs in the modern transportation sector. In effect, deployment of fast-charging stations for electrified transportation would enforce a fundamental shift in power system operation. The surge in charging demand from EVs can cause issues in reliable grid operation such as power losses and supply-demand mismatch, thereby causing grid-stability issues. Therefore, it is imperative to have smart control of aggregated EV charging demands, with pre-decided objectives of DR that take feedback from the grid operators or market mechanisms such that the power consumption profile of EVs is optimally adjusted in time and location. As discussed in detail in this chapter, the literature suggests that DR has significant potential in successful integration of electrified transportation sector with emerging renewable generation resources to ensure improved grid stability, economic welfare of DSO, EVAs, and EV users, whereas fulfilling auxiliary requirements of EV users such as battery degradation reduction or user privacy protection.

REFERENCES

[1] A. G. Anastasiadis, G. P. Kondylis, A. Polyzakis, and G. Vokas, "Effects of increased electric vehicles into a distribution network," *Energy Procedia*, vol. 157, pp. 586–593, 2019.

[2] NREL, "Incorporating Residential Smart Electric Vehicle Charging in Home Energy Management Systems." Available online at https://www.nrel.gov/docs/ fy21osti/78540.pdf.

[3] A. Fukushima, T. Yano, S. Imahara, H. Aisu, Y. Shimokawa, and Y. Shibata, "Prediction of energy consumption for new electric vehicle models by machine learning," *IET Intelligent Transport Systems*, vol. 12, no. 9, pp. 1174–1180, 2018.

[4] IEA, "Global EV Outlook 2022." Available online at https://www.iea.org/data-and-statistics/data-product/global-ev-outlook-2022.

[5] S. Lee and M. Park, "Understanding EV Market Trend: Using Time Series Dynamic Topic Modeling with Youtube Data," in *2021 IEEE International Conference on Big Data (Big Data)*, pp. 5941–5943, IEEE, 2021.

[6] H. Liimatainen, O. van Vliet, and D. Aplyn, "The potential of electric trucks–an inter- national commodity-level analysis," *Applied Energy*, vol. 236, pp. 804–814, 2019.

[7] ZeroKyst, "Decarbonising the fisheries and aquaculture industry." Available online at https://zerokyst.no/en/.

[8] U. B. Irshad, M. S. H. Nizami, S. Rafique, M. J. Hossain, and S. C. Mukhopadhyay, "A battery energy storage sizing method for parking lot equipped with EV chargers," *IEEE Systems Journal*, vol. 15, no. 3, pp. 4459–4469, 2020.

[9] M. Hosseinzadehtaher, D. Tiwari, N. Kouchakipour, A. Momeni, M. Lelic, and Z. Wu, "Grid Resilience Assessment during Extreme Fast Charging of Electric Vehicles via Developed Power Hardware-in-the-Loop," in *2022 IEEE Transportation Electrification Conference & Expo (ITEC)*, pp. 929–934, IEEE, 2022.

[10] J. Lee, H. Shon, I. Papakonstantinou, and S. Son, "Optimal fleet, battery, and charging infrastructure planning for reliable electric bus operations," *Transportation Research Part D: Transport and Environment*, vol. 100, p. 103066, 2021.

[11] Y. Alwesabi, Y. Wang, R. Avalos, and Z. Liu, "Electric bus scheduling under single depot dynamic wireless charging infrastructure planning," *Energy*, vol. 213, p. 118855, 2020.

[12] N. Gupta, "Probabilistic optimal reactive power planning with onshore and offshore wind generation, ev, and pv uncertainties," *IEEE Transactions on Industry Applications*, vol. 56, no. 4, pp. 4200–4213, 2020.

[13] V. Arzamasov, K. Bo¨hm, and P. Jochem, "Towards Concise Models of Grid Stability," in *2018 IEEE International Conference on Communications, Control, and Computing Technologies for Smart Grids (SmartGridComm)*, pp. 1–6, 2018.

[14] P. S. Kundur and O. P. Malik, *Power system stability and control.* McGraw-Hill Educa- tion, 2022.

[15] M. Nojavan and H. Seyedi, "Voltage stability constrained OPF in multi-micro-grid considering demand response programs," *IEEE Systems Journal*, vol. 14, no. 4, pp. 5221–5228, 2020.
[16] A. Primadianto and C.-N. Lu, "A review on distribution system state estimation," *IEEE Transactions on Power Systems*, vol. 32, no. 5, pp. 3875–3883, 2017.
[17] J. Zhao, Y. Wang, G. Song, P. Li, C. Wang, and J. Wu, "Congestion management method of low-voltage active distribution networks based on distribution locational marginal price," *IEEE Access*, vol. 7, pp. 32240–32255, 2019.
[18] J. Huang, B. Cui, X. Zhou, and A. Bernstein, "A generalized lindistflow model for power flow analysis," in *2021 60th IEEE Conference on Decision and Control (CDC)*, pp. 3493–3500, IEEE, 2021.
[19] H. Yuan, F. Li, Y. Wei, and J. Zhu, "Novel linearized power flow and linearized opf mod- els for active distribution networks with application in distribution LMP," *IEEE Transactions on Smart Grid*, vol. 9, no. 1, pp. 438–448, 2016.
[20] R. van Amerongen, "Sensitivity analysis of optimised power flows," *Archiv fu¨r Elek- trotechnik*, vol. 73, no. 1, pp. 59–67, 1990.
[21] H. S. Das, M. M. Rahman, S. Li, and C. Tan, "Electric vehicles standards, charging infrastructure, and impact on grid integration: A technological review," *Renewable and Sustainable Energy Reviews*, vol. 120, p. 109618, 2020.
[22] O. C. Alliance, "OPEN CHARGE POINT PROTOCOL 2.0.1." Available online at https://www.openchargealliance.org/protocols/ocpp-201/.
[23] G. of UK, "Residential chargepoints: minimum technical spec- ification." Available online at https://www.gov.uk/guidance/ residential-chargepoints-minimum-technical-specification.
[24] ELANDL, "Research: Smart Charging." Available online at https://www.elaad.nl/ research/ev-related-protocol-study/.
[25] W. Shen, T. T. Vo, and A. Kapoor, "Charging algorithms of lithium-ion batteries: An overview," in *2012 7th IEEE conference on industrial electronics and applications (ICIEA)*, pp. 1567–1572, IEEE, 2012.
[26] M. Restrepo, C. A. Can˜izares, and M. Kazerani, "Three-stage distribution feeder control considering four-quadrant EV chargers," *IEEE Transactions on Smart Grid*, vol. 9, no. 4, pp. 3736–3747, 2016.
[27] N. Mehboob, M. Restrepo, C. A. Canizares, C. Rosenberg, and M. Kazerani, "Smart operation of electric vehicles with four-quadrant chargers considering uncertainties," *IEEE Transactions on Smart Grid*, vol. 10, no. 3, pp. 2999–3009, 2018.
[28] H. Li, Z. Wan, and H. He, "Constrained EV charging scheduling based on safe deep reinforcement learning," *IEEE Transactions on Smart Grid*, vol. 11, no. 3, pp. 2427– 2439, 2019.
[29] C. S. Ioakimidis, D. Thomas, P. Rycerski, and K. N. Genikomsakis, "Peak shaving and valley filling of power consumption profile in non-residential buildings using an electric vehicle parking lot," *Energy*, vol. 148, pp. 148–158, 2018.

[30] Z. Wei, Y. Li, and L. Cai, "Electric vehicle charging scheme for a park-and-charge sys- tem considering battery degradation costs," *IEEE Transactions on Intelligent Vehicles*, vol. 3, no. 3, pp. 361–373, 2018.
[31] W. Infante, J. Ma, X. Han, and A. Liebman, "Optimal recourse strategy for battery swap- ping stations considering electric vehicle uncertainty," *IEEE Transactions on Intelligent Transportation Systems*, vol. 21, no. 4, pp. 1369–1379, 2019.
[32] A. Bracale, G. Carpinelli, and P. De Falco, "A probabilistic competitive ensemble method for short-term photovoltaic power forecasting," *IEEE Transactions on Sustainable Energy*, vol. 8, no. 2, pp. 551–560, Apr. 2017.
[33] R. T. Rockafellar and S. Uryasev, "Optimization of conditional value-at-risk," *Journal of Risk*, vol. 2, pp. 21–42, 2000.
[34] M. Ehrgott, J. Ide, and A. Scho¨bel, "Minmax robustness for multi-objective optimization problems," *European Journal of Operational Research*, vol. 239, no. 1, pp. 17–31, 2014.
[35] M. Yoshihara, T. Namerikawa, and Z. Qu, "Non-Cooperative optimization of Charging Scheduling of Electric Vehicle via Stackelberg game," in *2018 57th Annual Conference of the Society of Instrument and Control Engineers of Japan (SICE)*, pp. 1658–1663, IEEE, 2018.
[36] W. Tushar, T. K. Saha, C. Yuen, M. I. Azim, T. Morstyn, H. V. Poor, D. Niyato, and R. Bean, "A coalition formation game framework for peer-to-peer energy trading," *Applied Energy*, vol. 261, p. 114436, 2020.
[37] S. Boyd, N. Parikh, *et al.*, "Distributed optimization and statistical learning via the alter- nating direction method of multipliers," *Foundations and Trends in Machine learning*, vol. 3, no. 1, pp. 1–122, 2011.
[38] J. Yang, T. Wiedmann, F. Luo, G. Yan, F. Wen, and G. H. Broadbent, "A fully de- centralized hierarchical transactive energy framework for charging EVs with local DERs in power distribution systems," *IEEE Transactions on Transportation Electrification*, 2022.
[39] X. Zhang, M. M. Khalili, and M. Liu, "Improving the privacy and accuracy of ADMM-based distributed algorithms," in *International Conference on Machine Learning*, pp. 5796–5805, PMLR, 2018.
[40] F. Knirsch, D. Engel, C. Neureiter, M. Frincu, and V. Prasanna, "Model-driven privacy assessment in the smart grid," in *2015 International Conference on Information Systems Security and Privacy (ICISSP)*, pp. 1–9, 2015.
[41] M. Abouyoussef and M. Ismail, "Blockchain-based privacy-preserving networking strategy for dynamic wireless charging of EVs," *IEEE Transactions on Network and Service Management*, vol. 19, no. 2, pp. 1203–1215, 2022.

Chapter 7

Driver activity tracking for vehicle data recorder system

Vismaya M.K. and Trisiladevi C. Nagavi
JSS Science and Technology University Mysore

CONTENTS

DOI: 10.1201/9781003436089-7

7.1 INTRODUCTION

The digital tachograph is a truck-mounted device that keeps track of driving and resting times, speed and distance traveled. Additionally, it records the identities of the drivers, the tasks they carried out and any incidents that may have occurred such as speeding. Every truck must have a digital tachograph since 2006. The tachographs are typically used to ensure that drivers and employers adhere to the regulations regarding drivers' hours.

A digital tachograph keeps track of speed, distance, driving intervals, breaks and some incidents like speeding excessively and driving without a card. These data are primarily used to monitor drivers' compliance with rules governing their driving and resting periods, enabling businesses to uphold these rules and enforcers to conduct effective roadside inspections [1]. Drivers and their employers are legally obligated to accurately record their activities, keep the records and produce them upon request to transport authorities.

A digital driver card, the tachograph head and a sender unit mounted to the vehicle's gearbox make up a digital tachograph system. While the gearbox output shaft is turning, the sender unit generates electronic pulses. The head interprets these pulses as speed information. The pulses from the sender unit to the head are encrypted and electronically paired, preventing tampering by intercepting or duplicating the pulse signal in the intermediate wiring. The tachograph not only automatically receives speed data but also records the driver's activity in a mode of their choosing. When the vehicle is moving, the "drive mode" is activated automatically. Subsequently, when vehicle comes to a stop, digital tachograph heads typically default to the "other work" mode. The driver can manually switch between the "rest" and "availability" modes while parked. Drivers are required by law to accurately record their activities, keep the records and produce them to the transport authorities who are responsible for enforcing the laws governing drivers' working hours.

The proposed system adds a new feature in the vehicle data recorder system to make easier identification for a driver. That is driver identification is done via NFC. The system cost would be approximately Rs 10,000.

7.1.1 Existing and proposed system

In the existing system, the driver can be logged into the VDR system manually. In the proposed system, the driver can be identified through an NFC card and co-driver login and activities can be monitored.

7.1.2 Challenges and issues

Challenges and issues that are involved in the system development are as follows:

a. **Challenges**
   - The download time for data recorder could be faster, but it is too slow in some cases due to old transfer technology.
   - Unable to analyze driver without downloading tools and analysis software.
   - Driver login details may corrupt easily.

b. **Issues**
   - Increase the number of drivers inside the device memory. Nowadays, it is possible to register only up to 30 drivers inside the device memory.

7.1.3 Applications

Tachograph plays an important role in transport vehicles. Some of the applications where digital tachograph is being used are listed below:

a. A digital tachograph records data and allows for data analysis and cost forecasting.
b. Used in the transportation and mobility sectors where goods are transported via vehicles.
c. It enables employers to ensure that the Government's rules are followed at all times, from vehicle speed to total vehicle operating time.
d. It is one of the most important tools for truck drivers to use in order to manage their driving time.
e. It enables real-time asset and vehicle tracking and ensures driver safety.

7.2 LITERATURE REVIEW

It is observed that many research works in this area are based on near field communication (NFC), smart tachograph and vehicle data recorder. The following section discusses the literature based on these concepts where it helped to implement the proposed system.

7.2.1 Near field communication

The NFC is a subset of Radio Frequency Identification (RFID) technology. The modern age NFC was introduced in 2004 and has gained popularity since 2014, after ten years of development, owing to low-cost hardware, widespread use of smartphones and the boom in Internet of Things technology.

The NFC is a short-range wireless technology that allows mobile devices to interact with passive physical objects and many other active mobile devices, connecting the physical reality to mobile services in ways that empower and benefit users [2]. The RFID is a powerful enabling technology that is used in a wide range of applications and uses, including supply chain management and product inventory control as well as identity authentication and access control. As RFID technologies become more widely used, the possibility of unwanted identification, tracking and surveillance as well as data interception, cloning and misuse, may increase [3].

7.2.2 The smart tachograph

A system like the smart tachograph could be installed with little extra effort in modern vehicles. A modern car is equipped with a variety of sensors, including all of the sensors used in our prototype [4]. It has acceleration sensors for electronic stability programmers, a temperature sensor to warn the driver of a potentially slick road and light sensors for automatic headlight activation. Most mid- to high-end vehicles include a GPS sensor and navigation maps [5]. All of these sensors that have been installed in vehicles for other purposes could be reused for a system like the smart tachograph [6].

7.2.3 An in-vehicle data recorder (IVDR) for evaluation of driving behavior and safety

The overall framework and components of Drive Diagnostics, an IVDR and the results of a study to validate its performance are proposed by the authors [7]. The IVDR was created to monitor and analyze driver behavior in normal driving situations, not just crash or pre-crash events [8]. It records the vehicle's movement and uses this data to calculate the overall trip safety.

The purpose of this study is to assess the affectivity of a system designed to raise awareness and promote the procedure of accompanied driving. The study investigates the impact of accompanied driving on the performance of young drivers and other family members, as well as the issues of intergenerational behavior transfer.

7.2.4 Automated tachograph chart analysis system

This is an investigation into a new system for analyzing tachograph charts [9]. The circular charts are legal records that contain information on various types of driver activity like driving, other duty, standby and rest as well as vehicle data such as speed and distance traveled. Because every driver of a passenger or goods vehicle of a certain capacity is required to use a chart every 24 hours. Since tachograph charts are currently only examined manually, there is a clear need for automated analysis.

It is observed that many companies have adopted NFC-based technology for various products in this genre [10]. Here, we are developing contactless communication for users through an NFC card for the embedded system that is for vehicle data recorder.

7.3 METHODOLOGY

In a short context, a few technologies like NFC and RFID technology are briefly explained in the next subsection.

7.3.1 Core technologies

7.3.1.1 Near field communication

The NFC as its name suggests it is a shorter-range subset of RFID technology. It has gained popularity as the rising development technology in today's technical world. What this wireless communication technology offers are a low bandwidth with high frequency allowing data transfer in the range of centimeters.

Reading from NFC tags is very easy as we just need to bring NFC tag closer to the NFC reader and it will start reading from it without providing any connection details. The concept of inductive coupling is used in this study. It is also compatible with Bluetooth and Wi-Fi.

7.3.1.2 RFID Technology

The RFID emerged somewhere around in the 1980s. Charles Walton invented an object using RFID in 1983. It basically enables a one-way wireless communication that is typically between two devices, that is, a powerless RFID tag and a powered RFID reader. The RFID reader that is enabled with battery supply is responsible for generating long-distance radio frequency waves using which the RFID tag will get induced and generates its own electricity based on the strength of electromagnetic field received.

RFID can be scanned from a distance of 100 m without being in line of sight and that's why it is being used everywhere for asset tracking such as in a warehouse or airport and wild animal movement tracker or livestock identification.

The NFC works at high-frequency RFID band that is 13.56 MHz. The reason why this spectrum is accepted globally is because it is unlicensed and hence anyone can use it freely for transmitting and intercepting data.

7.3.2 Vehicle data recorder system design

The vehicle data recorder system is a digital electronic recording device, which records and stores data regarding the vehicle traveling speed, time, distance and related driving status information. All this information can be displayed through an interface USB stick or NFC card.

7.3.2.1 Context diagram

The pictorial representation of the context diagram in Figure 7.1 describes an overview of a system. The vehicle data recorder system can communicate with other vehicle-connected devices via interfaces. Driver can be identified by USB

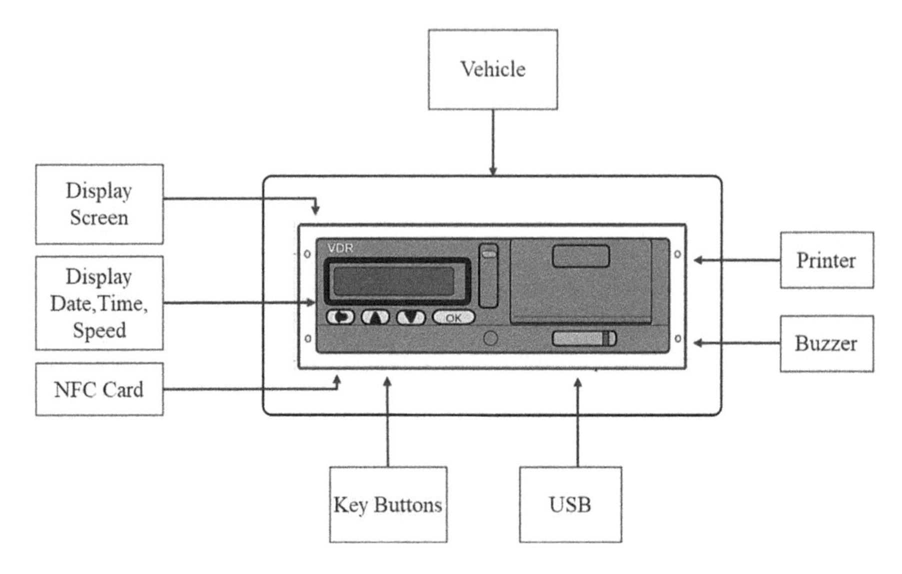

Figure 7.1 Context diagram of the vehicle data recorder system.

stick or NFC card or manual inputting identification code (via HMI). Driver activities are printed via its integrated thermal printer or data can be exported through USB from a device-generated file. Drivers can be able to navigate via its HMI menu and select certain functionalities using navigation buttons.

7.3.2.2 Vehicle data recorder offers the below interfaces to connect to the external environment

Buzzer– used to emit audible warnings.

HMI Buttons– used for navigation and triggering of various functions.

Display– used for showing various information to the users.

Printer– used to interface with a thermal printer for the purpose of printing various reports.

NFC– used to interface with the contactless cards for user identification (login/logout) and for work time control.

USB Interface– used to interface with USB stick for the user identification, software update, configuration update, data download and charging.

7.3.2.3 System requirements

The hardware and software requirements are listed below.

Hardware Requirements

- Device: BVDR 2.0 System, Controller area network protocol bus, RS232 protocol and speed sensor
- USB: 32/64 GB

- System: Intel i7, Workshop tab
- Others: Printing paper, cables to connect the device to initiate connection.

Software Requirements

- Operating System: Windows 7
- Coding Language: Embedded C
- IDE: Eclipse, doors, RFID Discover, CAN and K-line calibration tool, flashing tool, visual studio.

7.3.3 Vehicle data recorder system implementation

a. Use case diagram for driver activity using NFC

NFC– used to interact between vehicle data recorder system and contactless card for user identification (login/logout) and for work time control.

NFC is a set of communication protocols that enables communication between two electronic devices over 4 cm or less.

The main responsibility of NFC is as follows:

- Login/logout/swap of driver/co-driver
- Activity handling

In Figure 7.2, the use case diagram shall deal with the user's login to VDR and logout from VDR and fetches bytes from Library module whenever there is an NFC swipe in front of the device. NFC shall validate the Driver Code and Driver License Number by parsing the fetched bytes. If Driver Code and Driver License Number are valid, NFC shall inform User Manager Component about the User login. User Manager shall create the session for Driver/Co-driver.

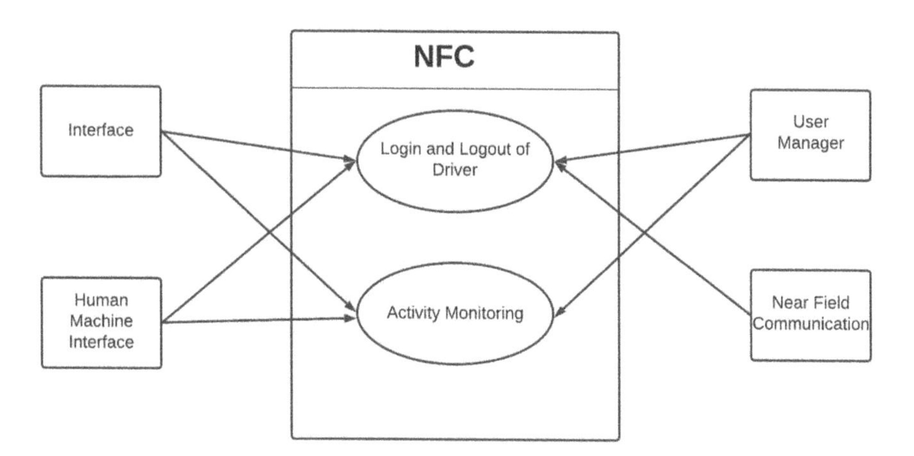

Figure 7.2 Use case diagram for NFC functionalities.

b. Activity diagram of login/logout of the user.

In Figure 7.3, the activity diagram represents the login and logout activities of a user. Firstly, the device should be in powered on which is considered mandatory precondition. Initially, the system mode is set to normal mode for tracking the further activity details of a user. Once this condition is satisfied, then the user can swipe his identification card where an NFC card starts read. Card can be preprogrammed with driver code and driver license number. If both credentials are valid and satisfy

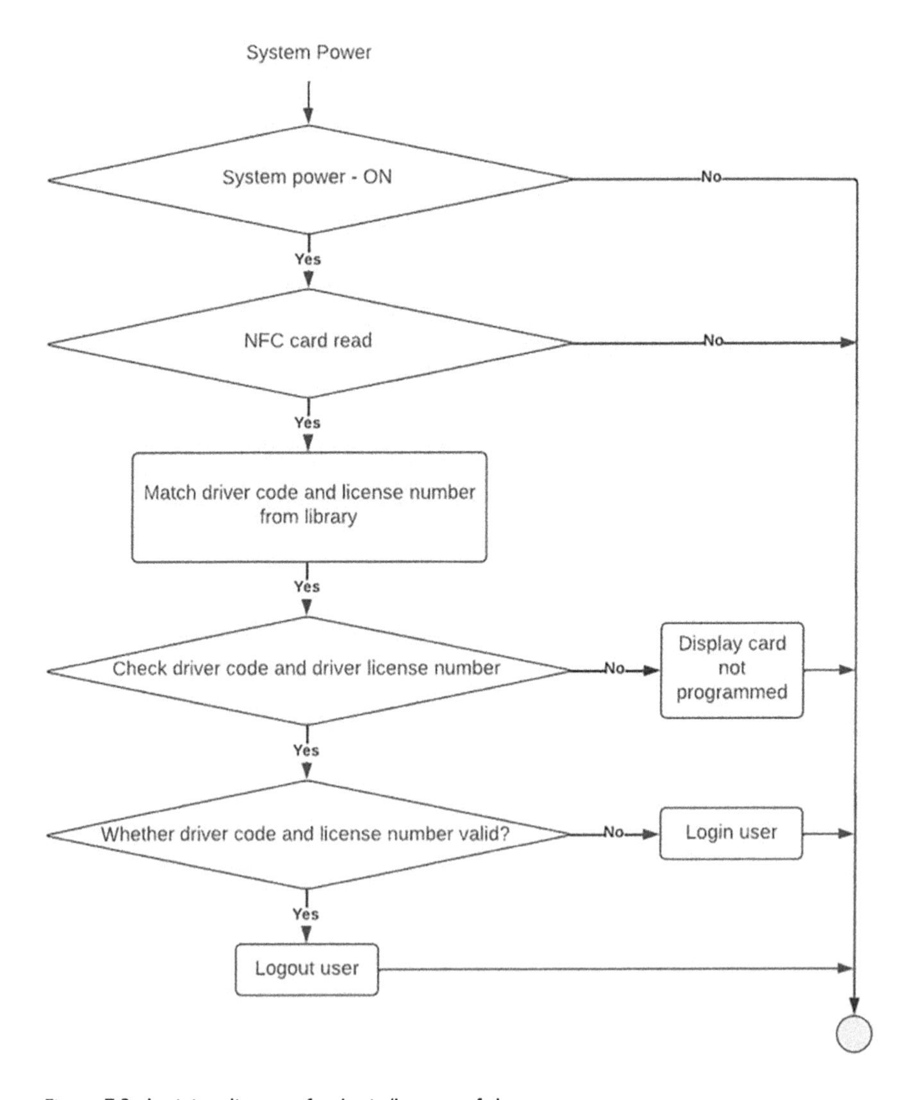

Figure 7.3 Activity diagram for login/logout of the user.

the condition and driving license number is not expired, then users can login to the system. If the user is with valid driver code and driving license number but card is expired then he won't be able to login to the device, it displays Driver License Number expired screen in a system. Swiping of card twice leads to logout of a driver as the same credentials are matching with the active driver session.

7.4 RESULTS ANALYSIS

In this section, the implementation of external and internal views of driver activity using an NFC card has been discussed.

7.4.1 View of driver activities in the system

The view of display screen when an NFC card is swiped to the system is discussed below.

a. **External view of driver activity when co-driveris disabled**
   The above screen in Figure 7.4 shows the driver login through an NFC card where driver code displays, here driver has been logged in with id 00001 through swiping an NFC card to a system.
b. **Internal view of driver activity when co-driver is disabled**
   Tracking information of a driver with an NFC card are discussed below.
   Once the driver is logged in, the details of speed distance, time of login and odometer details have been shown in Figure7.5.
   The above picture in Figure 7.6 shows the start time that is login time-andthe end time that is logout time with driver id and driver license number has been shown.
c. **External view of driver activity when co-driver is enabled**
   When the configuration is set to enable co-driver, then the user can login as a driver or co-driver using anNFC card.

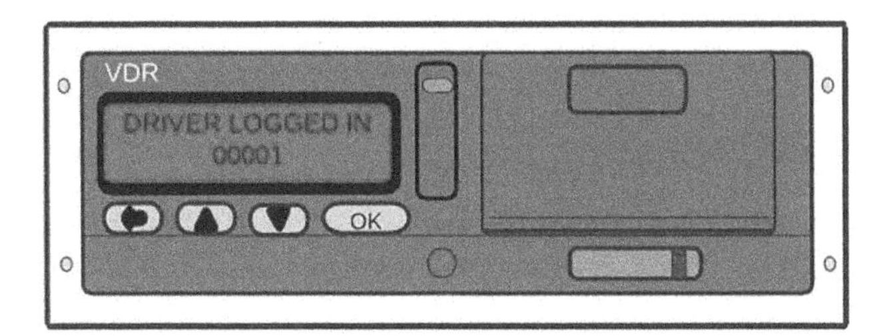

Figure 7.4 Driver login.

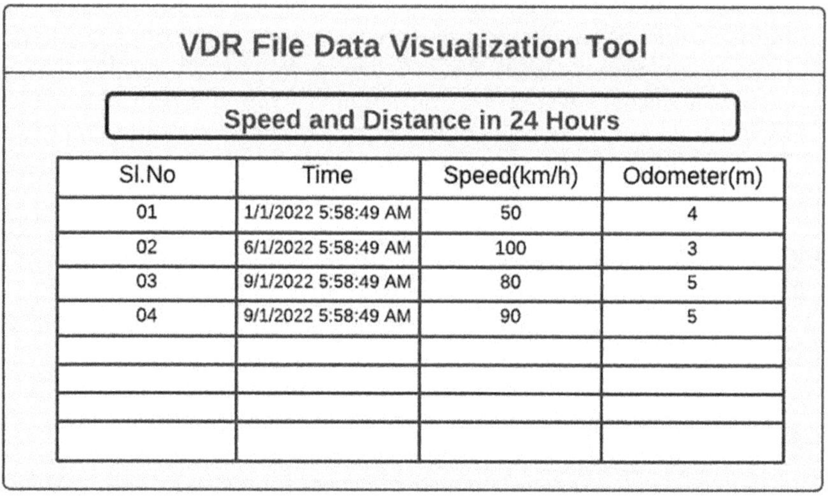

VDR File Data Visualization Tool

Speed and Distance in 24 Hours

Sl.No	Time	Speed(km/h)	Odometer(m)
01	1/1/2022 5:58:49 AM	50	4
02	6/1/2022 5:58:49 AM	100	3
03	9/1/2022 5:58:49 AM	80	5
04	9/1/2022 5:58:49 AM	90	5

Figure 7.5 Information of speed and distance in 24 hours.

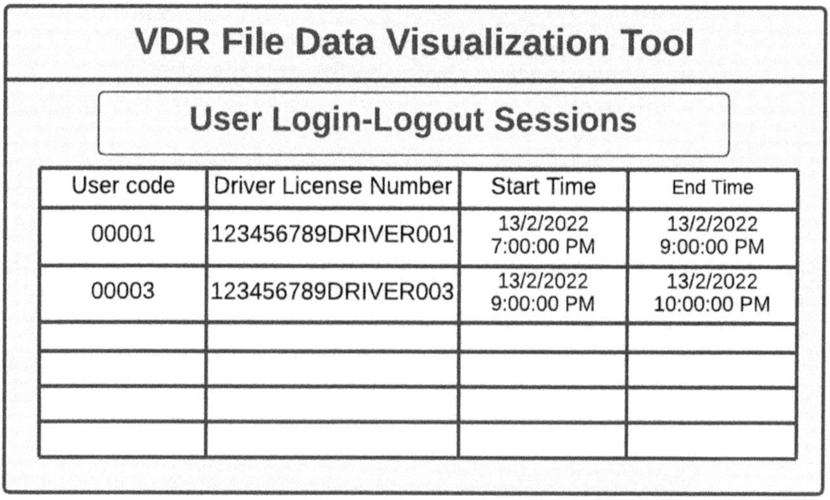

VDR File Data Visualization Tool

User Login-Logout Sessions

User code	Driver License Number	Start Time	End Time
00001	123456789DRIVER001	13/2/2022 7:00:00 PM	13/2/2022 9:00:00 PM
00003	123456789DRIVER003	13/2/2022 9:00:00 PM	13/2/2022 10:00:00 PM

Figure 7.6 User login–logout sessions.

When co-driver is enabled, the user can login to a system through card as a driver or co-driver after swiping card to a system is shown in Figure 7.7.

The above screen in Figure7.8 shows that the card holder with id 00003 as co-driver has been logged in to a system.

After swiping the same card twice, it displays whether to logout or to swap the driver and co-driver as shown in Figure 7.9.

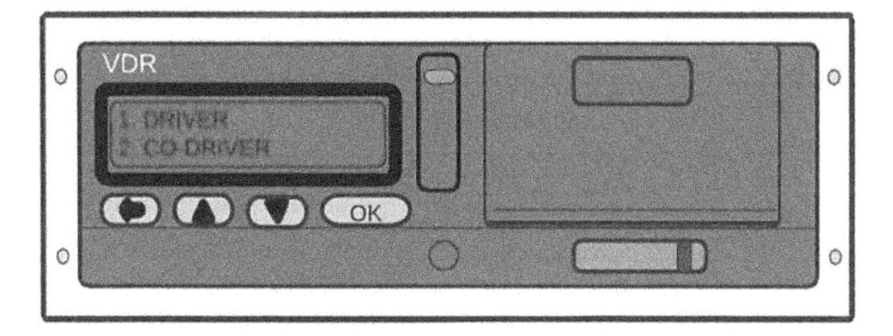

Figure 7.7 Login as driver or co-driver screen.

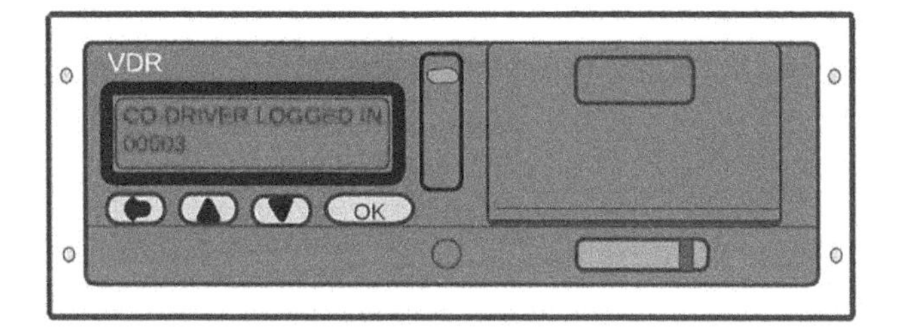

Figure 7.8 User logged in as co-driver after swiping card.

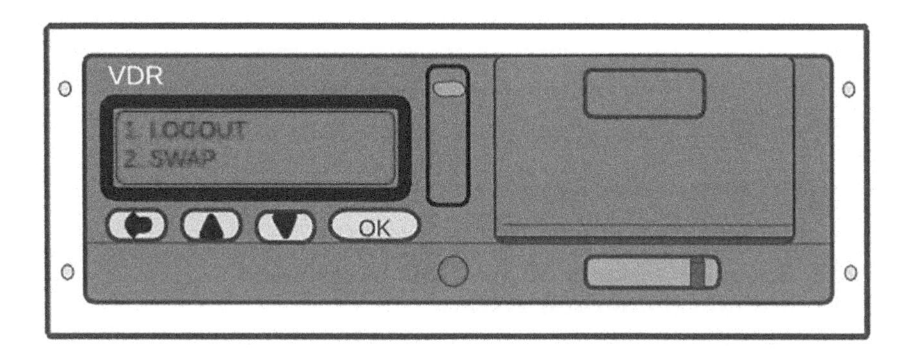

Figure 7.9 Logout or interchange position of driver and co-driver screen.

d. Internal view of driver activity when co-driver is enabled

The above screen inFigure7.10 shows the details of the current driver with driver license number, start time of the user logged in and the activity of the user.

The above screen inFigure 7.11shows the details of the current co-driver with driver license number and start time of the user logged in.

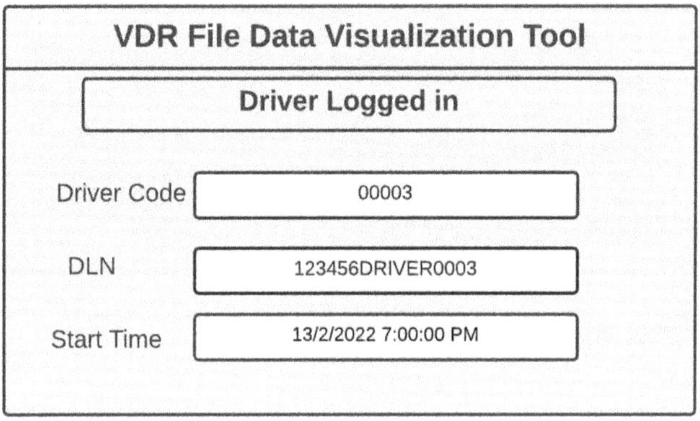

VDR File Data Visualization Tool

Driver Logged in

Driver Code 00003

DLN 123456DRIVER0003

Start Time 13/2/2022 7:00:00 PM

Figure 7.10 Current driver login details.

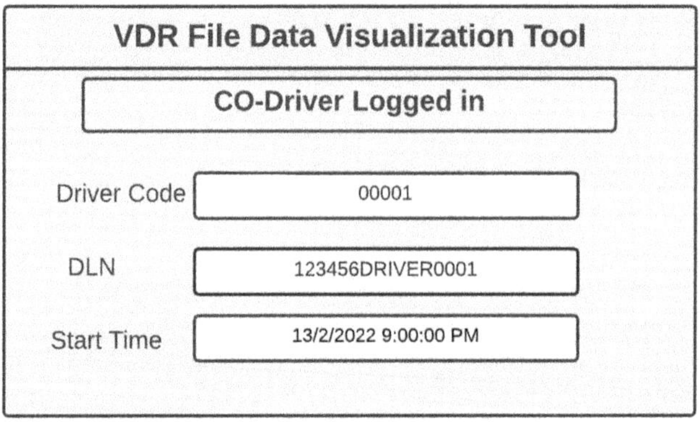

VDR File Data Visualization Tool

CO-Driver Logged in

Driver Code 00001

DLN 123456DRIVER0001

Start Time 13/2/2022 9:00:00 PM

Figure 7.11 Current co-driver login details.

7.4.2 Security aspects of the NFC card-based digital tachograph system

Initially, NFC card will be programmed by using an RFID Discover tool with driver code. While updating driver details to an NFC card, the data will be in the form of hexadecimal. Tracking information of a driver with an NFC card are exported through USB and using a VDR file data visualization tool. The activities can be viewed in a readable format. Each user will have a unique driver code and driver license number, hence data is secured and encrypted. For example,

3030303035A5A5- Driver ID →00005 (different for each user)

3132333435363742564452445249564552 31A5A5-LicenseNumber→1234567BVDRDRIVER1 (different for each user)

7.5 CONCLUSION AND FUTURE WORK

A tachograph records the driving, resting and service times as well as any potential irregularities based on the calibration index. To ensure that legal requirements are met and that road safety is improved, the data from the tachograph is the essential identifier. This chapter discusses driver communication using an NFC card, allowing for the recording of all relevant information about a driver's activity as well as the export of that information in the form of human-readable file using a USB. In the proposed system, it is observed that NFC technology can be implemented not only in electronic devices, but it can also be used for embedded systems for contactless communications.

By using NFC technology, contactless communication has become a safe and secure priority application to access many of the services. This technology provides an interface, which allows it to act as a smart card reader in electronic devices. A driver's log shows that they complied with the rules governing driving and service times and rest periods.

REFERENCES

[1] "Global Vehicle Traveling Data Recorder Market Growth 2021–2026", LP INFORMATION INC, October 2021.

[2] Ankit S. Sanghavi, Sagar D. Nikumb, Priyanka P. Bhoir and Sagar D. Pooja, RTO Automation System Using Near Field Communication (NFC), *International Journal of Advanced Research in Computer and Communication Engineering*, Vol. 6, Issue 3, pp. 763–765, March 2017. https://www.marketintelligencedata.com/categories/automotive-transportation

[3] Neeraj Singh, Near-Field Communication (NFC), *International Journal of Information Technology and Libraries*, Vol. 39, Issue 6, 2020. https://developer.android.com/guide/topics/connectivity/nfc.

[4] Mateusz Sudowski and Beata Mrugalska, Changing Data in Tachograph's Recording: A Case Study, *International Journal of Logistics and Transport*, Vol. 3, Issue 10, pp. 105–109, 2017. https://www.rsa.ie/services/professional-drivers/tachographs-information

[5] Marc Sel and Dusko Karaklajic, "Internet of Trucks and Digital Tachograph – Security and Privacy Threats," *Securing Electronic Business Processes*, pp. 230–238, Issue 10, Springer, 2014. https://www.thalesgroup.com/en/markets/digital-identity-and-security/government/driving-licence/digital-tachograph-card-project

[6] Vlad Coroama and Marc Langheinrich, "The Smart Tachograph," Video submission abstract. Adjunct Proceedings of UbiComp, Tokyo, Japan, September 2005.

[7] Tsippy Lotan and Tomer Toledo, "An In-Vehicle Data Recorder for Evaluation of Driving Behavior and Safety", *Safety Data, Analysis, and Evaluation*, pp. 112–119, Transportation Research Board, 2006.

[8] Dóra Horváth and Roland Zs. Szabó, Driving Forces and Barriers of Industry 4.0: Do Multinational and Small and Medium-Sized Companies Have Equal Opportunities?, *International Journal of Technological Forecasting*

and Social Change, Vol. 146, pp. 119–132, 2019. https://soccernurds.com/uncategorized/2779493/exclusive-research-on-digital-tachographs-market-2021-key-players-industry-insight-growth-driver-analysis/

[9] A. Antonacopoulos and D.P. Kennedy, "An In-Vehicle Data Recorder for Evaluation of Driving Behavior and Safety", 2002.

[10] Hongjiang He and Yamin Zhang, Research on Vehicle Traveling Data Recorder, in *Proceedings of 2nd IEEE International Conference on Intelligent Computation Technology and Automation*, Vol. 2, pp. 736–738, November 2019. https://www.aeresearch.net/vehicle-traveling-data-recorder-market-700489.

Chapter 8

Robust control of hybrid maglev-based transportation system

Prince Kumar Saini and Bhanu Pratap
National Institute of Technology Kurukshetra

CONTENTS

8.1 INTRODUCTION

The transportation system is one of the most significant factors for a country's progress. Although this sector has a large and diverse scope with its own share of challenges, they can be overcome by the latest-efficient technologies. Magnetic levitation (Maglev) is one of the best-suited transportation for next-generation transportation systems, with the advantages of high speed, environmental responsiveness, and low noise. The hybrid maglev transportation system (HMTS) can be broadly classified as (Hu et al., 2021): (i) electromagnetic suspension (EMS) systems and (ii) electrodynamic suspension (EDS) systems.

The majority of maglev systems worldwide use EMS systems, which are based on active feedback control of the air gap to maintain stable levitation. EMS maglev uses powered electromagnets that are attracted

DOI: 10.1201/9781003436089-8

to an iron rail on the track. They usually operate with a small air gap and require accurate track construction for stable levitation. EMS maglev has good ride quality and is ideally suited for use. However, EDS systems produce repulsive force between levitation and guideway magnets. This repulsive force is enough to overcome the gravitational force and allows it to levitate. The EDS maglev requires a forward motion for levitation but no active control for stable levitation. These systems produce large air gaps and are inherently safe in the case of loss of power. Since the drag forces acting on the maglev vehicle are reduced as speed increases, EDS is ideally suited for very high speeds, and using superconducting material coils makes them expensive. Only a few examples of the EDS maglev that have been built or tested are mentioned by Hu et al. (2021). The levitation forces are the key ideology of maglev-based transportation systems. Simple coils produce these forces in conventional levitation systems. Hence, a large amount of heat loss in the coils degrades the performance and results in higher maintenance costs. So due to these reasons, a hybrid levitation system is needed.

The HMTS is classified into (i) permanent magnet and electromagnet hybrid, (ii) high-temperature superconductor (HTS), and (iii) normal operating electromagnet hybrid. The primary levitation forces produced by permanent magnets are used to control the guided pathway in the electromagnet-based HLS. The latter uses normal conducting coils and HMTS coils to make hybrid levitation electromagnets. A similar demonstration that focuses on an electromagnet designed with a high-superconducting coil (HTS) with a vehicle running speed of 500 km/h was investigated by Lee (2011). The hybrid coils in HLS need a constant current supply to give the required suspension force for the desired levitation, and the normal coil is required to control the conduction in HTS, so they are also regarded as control coils. This reduces the energy consumption for HLS, so they mostly operate at zero power modes.

The book chapter focuses on the application of robust control for the magnetic levitation system used in maglev trains. The earlier approach for maglev control is based on linearization and approximation methods at an equilibrium point and then the application of well-known controllers such as proportional-integral-differential and linear-quadratic regulator, proposed by Ding et al. (2018) and Yang et al. (2010), respectively. These conventional methods degrade the performance and are less robust. The feedback linearization technique has been applied to control a maglev system (Hajjaji et al. 2001). It uses mapping techniques and nonlinear transformation to transform a nonlinear dynamic model of a system into a linear model. It is an effective method to tackle many control applications and has been applied by many researchers across the globe for HMTS (Ding et al. 2018).

Recent studies used the backstepping technique to model the uncertainties for robust control application to the maglev-based system (Lee et al. 2011). Also, more advanced controllers such as adaptive control, sliding control, and a combination of both have been applied to maglev-based trains. An observer-based robust controller is designed by Ni et al. (2021) to enhance the suspension force model with the effect of disturbance. Xu et al. (2018) have also formulated an observer-based adaptive sliding mode control (SMC) law for maglev trains, but insensitive to external disturbances and uncertainty. Intelligent control methods are developed to deal with the complex nonlinearity of maglev systems. Su et al. (2014) proposed a hybrid robust control law based on the fuzzy model for the EMS system. This novice method handled the parameter uncertainties and improved the robustness.

The combined SMC with backstepping control approach is presented by Yang and Yen (2018) to tackle uncertain strict-feedback nonlinear systems. Sacchi et al. (2022) proposed a novel neural network-based practical SMC design for a nonlinear system capable of approximating unknown terms. Two deep neural networks construct the neural network-based implementation for predicting and estimating unknown parameters. Sun et al. (2022) proposed an amplitude saturation controller for EMS-type maglev vehicles to keep the proper air gap between the electromagnet and the rail track. The RBFNN-based supervisor controller is designed to improve track flexibility, force saturation, and the feedback signal. Chen et al. (2021) developed a neuro-state observer to estimate system states and parameters and used it to design an inverse control law. The simulation results prove the fast response and stability. An onboard image detection system has been designed by He et al. (2022) for high-speed maglev tracks using deep learning. These control strategies for the maglev system deal with external disturbances and model uncertainty; however, most researchers have not considered the parameter constraints such as air gap, acceleration, maximum control current, pole area, and leakage flux. These constraints are essential in realizing a robust and flexible maglev-based transportation system.

In this chapter, the development of robust control for HMTS has been aimed at considering system constraints. The proposed controller has been designed using a sliding-backstepping control scheme to stabilize HMTS. Also, the chattering problem has been taken into account by incorporating the adaptive gain scheduling (AGS) approach based on RBFNN to attain the robust performance of HMTS in the presence of parametric uncertainties and external disturbances. The rest of the chapter is organized as follows. In Section 8.2, the dynamical modeling of HTMS is illustrated. The design of a robust controller for HMTS is presented in Section 8.3. The simulation study has been carried out in Section 8.5. Finally, conclusions are given in Section 8.6.

8.2 DYNAMICAL MODELING OF HMTS

Numerous maglev transportation systems have been constructed, tested, and improved in recent years. Generally, an HMTS mechanism is divided into (i) propulsion mechanism and (ii) levitation mechanism. It has been used widely in industrial processes due to its advantageous performance features such as high-starting thrust force, high-speed operation, reduction of mechanical losses, alleviation of gear between the motor and the motion devices, and the size of motion devices. In the HMTS, magnetic levitation (maglev) technology has been manipulated to eliminate friction due to mechanical contact, decrease cost, and achieve high-precision positioning.

The HMTS is made up of a controller, an F-shaped rail, and a hybrid electromagnet. The U-shaped iron core electromagnet is generally used with conducting coils for levitation purposes. The levitation module is made up of two or more independent hybrid electromagnets, which is known as a single-point hybrid levitation system. This module is used as a control object for EMS-based maglev trains. The simple hybrid levitation system is shown in Figure 8.1.

The magnetic Equivalent Circuit Model estimates the amount of magnetic flux associated with a hybrid levitation system.

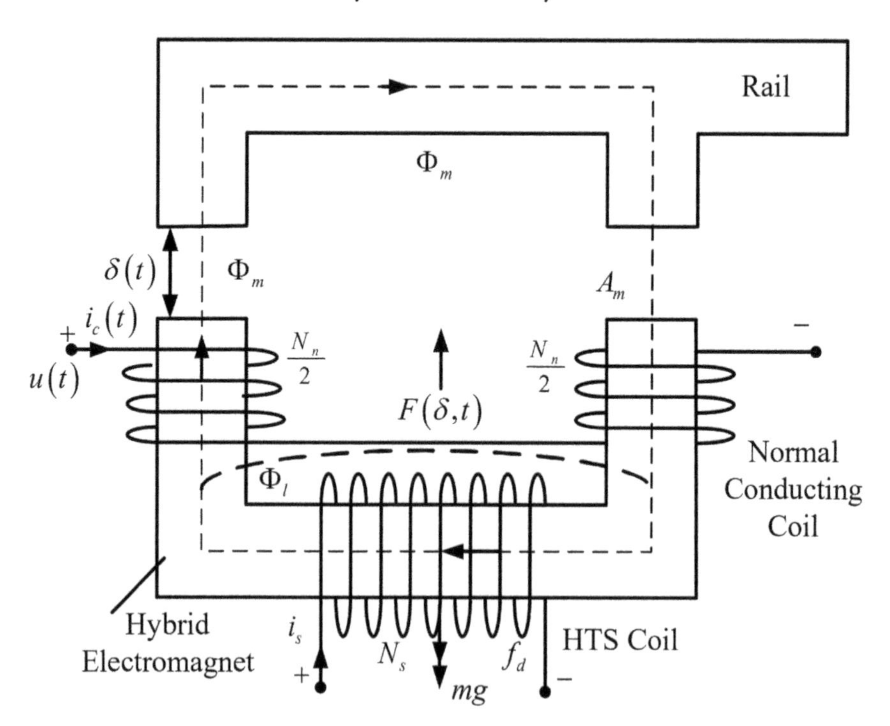

Figure 8.1 Schematic diagram of the hybrid levitation system (Hu et al., 2021).

In this magnetic reluctance, rail and iron core are neglected, and elastic vibration is not considered.

The magnetic flux Φ_m can be obtained using Hopkinson's law (Hu et al., 2021), given by

$$\Phi_m(\delta, i_c) = \frac{\mu_o A_m (N_n i_c + N_s i_s)}{2\delta} \tag{8.1}$$

where μ_o is the vacuum permeability, A_m is the cross-sectional area, δ is the air gap between the rail and hybrid electromagnet, N_n and N_s are the numbers of winding turns of the normal conducting coil and HTS coil, and i_c and i_s are the coil current and constant current, respectively.

At the airgap δ, the magnetic flux density B can be expressed as

$$B = \frac{\mu_o A_m (N_n i_c + N_s i_s)}{2\delta} \tag{8.2}$$

The magnetic levitation force F can be formulated by the Maxwell stress tensor method, given by

$$F(\delta, i_c) = \mu_o A_m \left[\frac{(N_n i_c + N_s i_s)}{2\delta} \right]^2 \tag{8.3}$$

Hybrid Levitation System Model: The equation of normal conducting coils using Kirchhoff's voltage law is given as

$$V_c = R i_c + L \frac{di_c}{dt} - \frac{\mu_o N_n A_m (N_n i_c + N_s i_s)}{2\delta^2} \frac{d\delta}{dt} \tag{8.4}$$

where R and L denote the electric resistance and the inductance of normal conducting coils, respectively. The inductance $L = \dfrac{\mu_0 N_n^2 A_m}{2\delta(t)}$.

The equation of hybrid electromagnet using Newton's Second Law in the vertical direction is represented as

$$m \frac{d^2\delta}{dt^2} = mg - F(\delta, i_c) + f_d \tag{8.5}$$

where m is the mass of the hybrid electromagnet, g is the acceleration due to gravity, and f_d is the external disturbance.

8.2.1 Nonlinear state-space model of HMTS

Combining the above equations, the nonlinear model of HMTS can be obtained as

$$\begin{aligned} \dot{x} &= f(x) + g(x)u + E_d \\ y &= h(x) \end{aligned} \tag{8.6}$$

where

$x = \begin{bmatrix} \delta & \dot{\delta} & i_c \end{bmatrix}^T$ is the state vector,
$u = V_c$ is the control input,
$y = \delta$ is the output.
The nonlinear terms $f(x)$, $g(x)$, and $h(x)$ are given by

$$f(x) = \begin{bmatrix} x_2 \\ g - \frac{K_s}{mx_1^2}\left\{x_3 + \frac{N_s i_s}{N_n}\right\}^2 \\ \frac{x_2 x_3}{x_1} + \frac{N_s i_s x_2}{N_n x_1} - \frac{R x_1 x_3}{2K_s} \end{bmatrix}, g(x) = \frac{x_1}{2K_s}, E_d = \frac{f_d}{m}, h(x) = x_1, \text{ and}$$

$$K_s = \frac{\mu_o N_n{}^2 A_m}{4}.$$

8.2.2 Nonlinear model of HMTS in normal form

This section develops a stable nonlinear model of HTMS in normal form. They are differencing the output results of the robust relative degree of HMTS (8.6) as 3. The normal form of the HMTS plant can be obtained using the Lie derivative of the function $h(x)$ upto third-order differential.

$$X = \begin{bmatrix} y \\ \dot{y} \\ \ddot{y} \end{bmatrix} = \begin{bmatrix} L_f^0 h(x) \\ L_f^1 h(x) \\ L_f^2 h(x) \end{bmatrix} \equiv \begin{bmatrix} X_1 \\ X_2 \\ X_3 \end{bmatrix}$$

Thus, the HMTS model in normal form can be expressed as

$$\dot{X} = \begin{bmatrix} \dot{X}_1 \\ \dot{X}_2 \\ \dot{X}_3 \end{bmatrix} = \begin{bmatrix} L_f^1 h(x) \\ L_f^2 h(x) \\ L_f^3 h(x) + L_g L_f^2 h(x) u \end{bmatrix} \equiv \begin{bmatrix} X_2 \\ X_3 \\ f_X + g_X u \end{bmatrix} \tag{8.7}$$

where $f_X = \dfrac{Rx_3}{mx_1}\left\{x_3 + \dfrac{N_s i_s}{N_n}\right\}$ and $g_X = -\dfrac{N_n x_3 + N_s i_s}{mN_n x_1}$.

8.2.3 Control objective

The objective of this chapter is to present a robust control scheme for an uncertain model of HMTS to attain the following tasks:

i. the plant output track the desired signal quickly,
ii. the tracking error approaches zero asymptotically,
iii. the control input remains under system constraints.

8.3 DESIGN OF ROBUST CONTROLLER FOR HMTS

This section has developed a robust control scheme for HMTS using the backstepping sliding mode control (BSMC) technique. The BSMC is one of the most effective control techniques for systems having nonlinear dynamics in the strict-feedback form. The BSMC is a step-by-step recursive technique stabilizing the system states and reducing design complexity. The theory of Lyapunov guarantees the stability of the closed-loop HMTS.

The SMC is a renowned robust control scheme that has been effectively and commonly applied for linear and nonlinear systems. Although the linear sliding surface parameters can be adjusted appropriately to obtain the arbitrary convergence rate, the system states cannot reach equilibrium in a finite time. ETC-based SMC is recently proposed to overcome this drawback based on the concept event triggering rule. Compared with the conventional SMC, it offers some superior properties such as faster, finite-time convergence and better control precision. SMC is a nonlinear control method that alters the dynamics of a nonlinear system by applying a discontinuous control signal that forces the system to "slide" along a cross-section of the system's normal behavior. The state-feedback control law is not a continuous function of time. Instead, it can switch from one continuous structure to another based on the current position in the state space. Hence, sliding mode control is a variable structure control method. The multiple control structures are designed so that trajectories always move toward an adjacent region with a different control structure. So, the ultimate trajectory will not exist entirely within one control structure. Instead, it will slide along the boundaries of the control structures.

The backstepping controller design methodology provides an effective tool for designing controllers for a large class of nonlinear systems with a triangular structure. The basic idea behind backstepping is to break a design problem on the entire system down to a sequence of sub-problems on lower order systems and recursively use some states as "virtual controls" to obtain the intermediate control laws with the Control Lyapunov Function. Starting from a lower order system and dealing with the interaction after augmentation of new dynamics makes the controller design easy. The advantages of backstepping control include guaranteed global and regional stability, the stress on robustness, and computable transient performance. The backstepping method has received great interest since its proposition and has been widely applied to control aerospace and mechanical engineering problems, etc. Along with these years of studies, this method has evolved to be relatively systematic and inclusive. For example, techniques like nonlinear damping, variable structure control, neural network adaptive control, and fuzzy adaptive control are synthesized to address various uncertainties, including matching and un-matching.

Consider the nonlinear model of HMTS (8.8) in the normal form with uncertainty,

$$\begin{aligned} \dot{X}_1 &= X_2 \\ \dot{X}_2 &= X_3 \\ \dot{X}_3 &= f_X + g_X u + E_{Xd} \\ y &= X_1 \end{aligned} \tag{8.8}$$

Assumption-1: It is assumed that y_r the desired signal and its derivatives up to the third order are bounded with a known positive constant Y such that $\|y_r\| \le Y$.

Assumption-2: The uncertainty present in the HMTS plant is assumed to be bounded such that $\|E_{Xd}\| \le E$.

Algorithm Design

INPUTS

i. Referenced output y_r
ii. HMTS plant output y

OUTPUTS

Step 1: (Design of virtual control law α_1)

a. Assume the first tracking error $e_1 = y - y_r$
b. The first error dynamics $\dot{e}_1 = X_2 - \dot{y}_r$ (8.9)
c. Choose the virtual control law $\alpha_1 = -K_1 e_1 + \dot{y}_r$; where $k_1 > 0$ (8.10)
d. Calculate the derivative of e_1 as $\dot{e}_1 = -K_1 e_1 + e_2$ (8.11)

Step 2: (Design of virtual control law α_2)

a. Assume the second tracking error $e_2 = X_2 - \alpha_1$
b. The second error dynamics $\dot{e}_2 = X_3 - \dot{\alpha}_1$ (8.12)

c. Choose the virtual control law $\alpha_2 = -K_2 e_2 + \dot{\alpha}_1 - e_1$; where $k_2 > 0$ (8.13)

d. Calculate the derivative of e_2 as $\dot{e}_2 = -K_2 e_2 + e_3 - e_1$ (8.14)

Step 3: (Design of the actual control law u)

a. Assume the third tracking error $e_3 = X_3 - \alpha_2$

b. Define the sliding surface $s = ce_2 + e_3$; where $c > 0$ (8.15)

c. Derivative of the sliding surface

$$\dot{s} = c(-K_2 e_2 + e_3 - e_1) + f_X + g_X u + E_{Xd} - \dot{\alpha}_2 \tag{8.16}$$

where $c > 0$

d. Choose the actual control law

$$u = \frac{1}{g_X}\begin{bmatrix} -f_X + \dot{\alpha}_2 - c \\ (-K_2 e_2 + e_3 - e_1) \\ -e_2 + u_r \end{bmatrix} \tag{8.17}$$

e. where $u_r = -\varsigma s - K\,\mathrm{sgn}(s)$, $\varsigma > 0$, $K > 0$ (8.18)

f. Calculate the derivative of s as $\dot{s} = -e_2 + u_r + E_{Xd}$ (8.19)

The above algorithm is terminated if the error is under the acceptance limit; otherwise, start from step 1. The Lyapunov-based approach analyzes the stability of the proposed control scheme for the HMTS. Considering a Lyapunov function,

$$V_H = \frac{1}{2}e_1^2 + \frac{1}{2}e_2^2 + \frac{1}{2}s^2 \tag{8.20}$$

Differentiating (8.20) as

$$\dot{V}_H = e_1\dot{e}_2 + e_2\dot{e}_2 + s\dot{s}.. \tag{8.21}$$

and substituting (8.11), (8.14), and (8.19) results

$$\dot{V}_H = e_1(-K_1 e_1 + e_2) + e_2(-K_2 e_2 + e_3 - e_1) + s(-e_2 + u_r + E_{Xd})$$

$$\dot{V}_H = -K_1 e_1^2 + e_1 e_2 - K_2 e_2^2 + e_2 e_3 - e_1 e_2 - se_2 + su_r + sE_{Xd} \tag{8.22}$$

Substituting (8.16) and (8.18) into (8.22) gives

$$\dot{V}_H = -K_1 e_1^2 - K_2 e_2^2 + e_2 e_3 - (ce_2 + e_3)e_2 + s\{-\varsigma s - K\,\mathrm{sgn}(s)\} + sE_{Xd}$$

$$\dot{V}_H = -K_1 e_1^2 - K_2 e_2^2 + e_2 e_3 - ce_2^2 - e_2 e_3 - \varsigma s^2 - Ks\,\mathrm{sgn}(s) + sE_{Xd} \tag{8.23}$$

Applying Assumption-2 in (8.23)

$$\dot{V}_H \leq -K_1 e_1^2 - K_2 e_2^2 - c e_2^2 - \varsigma s^2 - (K - E)|s| \tag{8.24}$$

By choosing the gain $K > E$ such that expression (8.23) can be rewritten as

$$\dot{V}_H \leq -K_1 e_1^2 - K_2 e_2^2 - c e_2^2 - \varsigma s^2 \tag{8.25}$$

From the above expression, it is obvious that all the error signals and sliding surfaces are bounded, which ensures the stability of the proposed control scheme for HMTS.

8.4 ROBUST CONTROL USING AGS BASED ON NEURAL NETWORK

The proposed scheme for HMTS provides an effective, robust control approach. However, chattering is a common phenomenon to be avoided by providing a smooth control action with RBFNN-based AGS. The inputs to the RBFNN are the sliding surface with its differential, whereas the output is gain K.

Algorithm Design

INPUTS

a. sliding surface s
b. its differential $\dot{s}$

OUTPUT

The gain K_N of the sign function

Step 1: Defining the input layer

The sliding surface and its differential are inputs to the RBFNN as $\bar{s} = \begin{bmatrix} s & \dot{s} \end{bmatrix}^T$.

Step 2: Selecting the membership function given as

$$\phi(\bar{s}) = \exp\left\{\frac{\|\bar{s} - \mu_i\|^2}{\sigma_i^2}\right\} \tag{8.26}$$

where mean and the standard deviations are represented by μ and σ_i, respectively.

Step 3: Calculating the output gain of the sgn function

$$K_n = w^T \phi(\bar{s}) \tag{8.27}$$

where w is the neural network weight.

Step 4: Design of learning algorithm

a. Select the energy function $E = \frac{1}{2}e^2$

   where the tracking error $e = y - y_r$

b. Design the learning algorithm based on the back-propagation method:

$$\Delta w \approx -\gamma e\left(-g_X^{-1} \operatorname{sgn}(\bar{s})\phi(\bar{s})\right)\operatorname{sgn}\left(w^T\phi(\bar{s})\right) \tag{8.28}$$

where γ is the learning rate of RBFNN.

Based on the above learning algorithm of RBFNN, the proposed control law is given by

$$u_N = \frac{1}{g_X}\left[-f_X + \dot{\alpha}_2 - c(-K_2e_2 + e_3 - e_1) - e_2 - \varsigma s - K_n \operatorname{sgn}(s)\right] \tag{8.29}$$

The proposed robust controller was designed for HMTS using an RBFNN-based AGS scheme, which is demonstrated with the help of a block diagram (Figure 8.2). The performance of a closed-loop HMTS has been analyzed in the next section.

8.5 SIMULATION STUDY

The simulation study has been conducted to verify the efficacy of the proposed robust controller for HMTS. The proposed control scheme provides an effective control action for the HMTS. The proposed control law includes he sign function, which results in the sliding surface's repeated switching, and the signum function's gain determines the chattering intensity. An RBFNN is adopted to construct the adaptive control

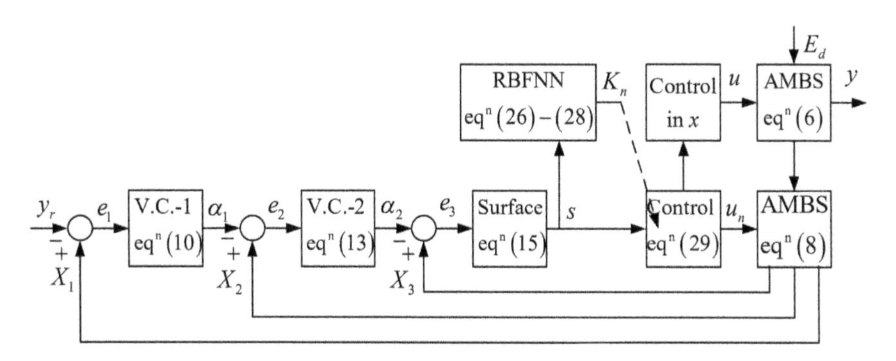

Figure 8.2 Block diagram of NNAGS-BSMC [11].

Table 8.1 Parameters of the HMTS

HMTS parameters	
Parameters	*Values*
m	311.84 kg
g	9.81 ms^2
R	1 Ω
N_n	500
N_s	594
i_s	24.75 A
μ_0	$4\pi \times 10^{-7}$ H/m
A_m	0.018 m^2

Table 8.2 Parameters proposed controller

Controller parameters	
Parameters	*Values*
K_1	25
K_2	2
c	20
ς	5
γ	0.1
K_n	Adaptive Gain
y_r	$0.5\sin(0.3t)$
x_0	$\begin{bmatrix} 0.1 & 0 & 0 \end{bmatrix}^T$

on the same control law to address this problem. The initial weights of the neural network are selected as zeros. The inputs to RBFNN are s and $\dot{s}$. This chapter has successfully demonstrated the development and application of the robust and adaptive control law for uncertain HMTS with parametric uncertainties and external disturbances. The simulation results have been obtained with the chosen parameters of the HMTS and proposed controller.

8.5.1 Simulation results

The simulation results reveal the effectiveness of the proposed robust control designed for the HMTS in the above plots. The tracking of output (measured air gap) with the desired trajectory is shown in Figure 8.3. This shows the fast convergence of the measured air gap to the desired. The output tracking error tends to zero quickly, as given in Figure 8.4. Similarly, the tracking of measured electromagnet velocity with desired is shown in Figure 8.5. The measured velocity converges to the desired velocity with the least amount of error. The electromagnet velocity tracking error tends to zero, given in Figure 8.6. The coil current is plotted in Figure 8.7. The coil current plays an essential role in HMTS as it acts as a bias indicator for accurate tracking. In contrast, the sliding surface of the robust control scheme is illustrated in Figure 8.8. The results are extended with the use of robust control law with RBFNN to consolidate the effectiveness of HMTS for more robustness. The controller gain used in the proposed control law with the signum function is updated using RBFNN and depicted in Figure 8.9. The voltage, as the control effort given to the HMTS has demonstrated in Figure 8.10, is smooth

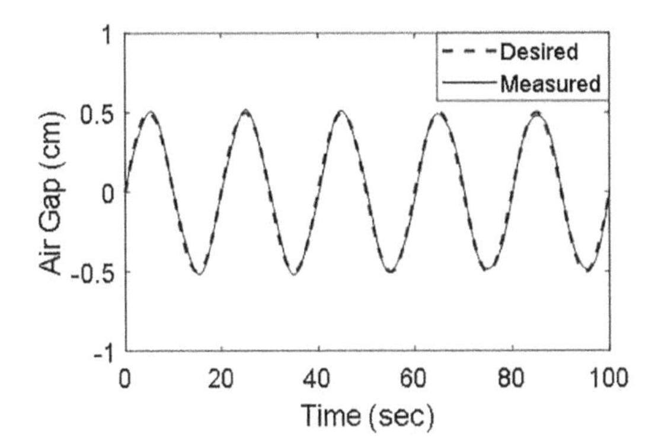

Figure 8.3 Tracking of air gap (cm).

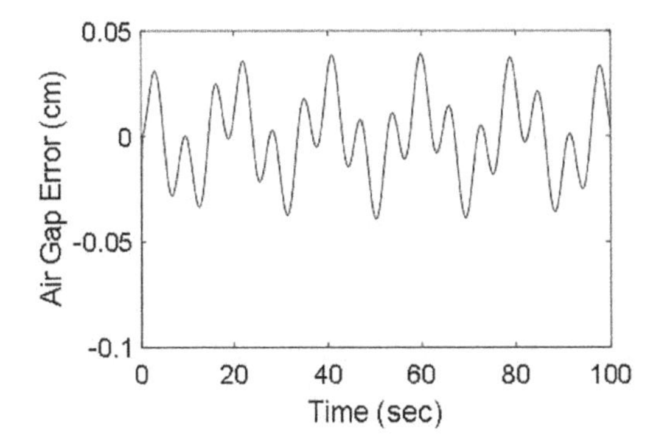

Figure 8.4 Tracking error of air gap (cm).

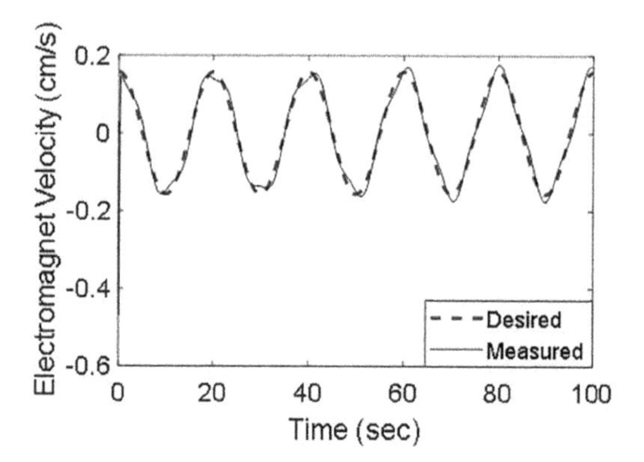

Figure 8.5 Tracking of electromagnet velocity (cm/s).

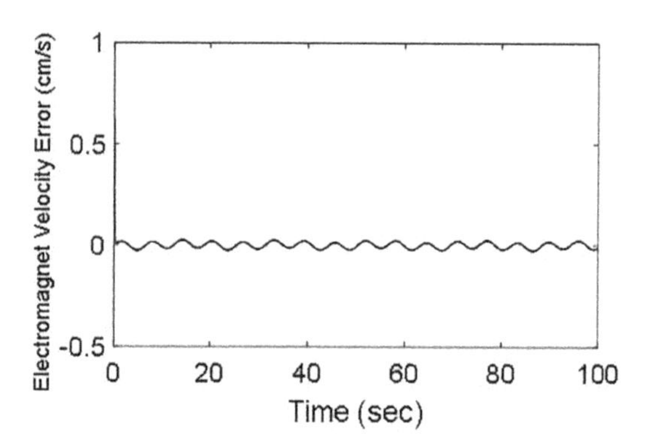

Figure 8.6 Tracking error of electromagnet velocity (cm/s).

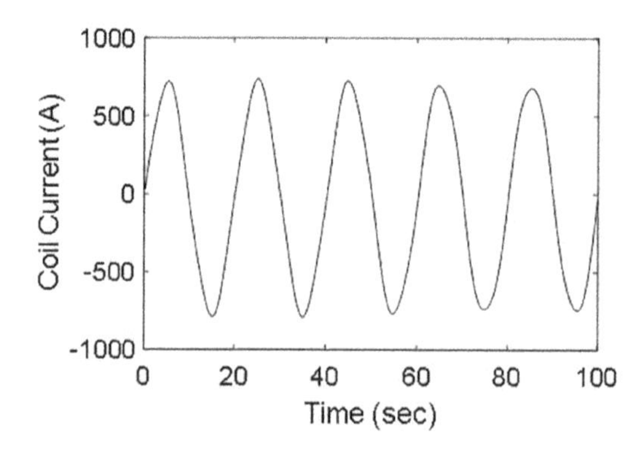

Figure 8.7 Coil current (A).

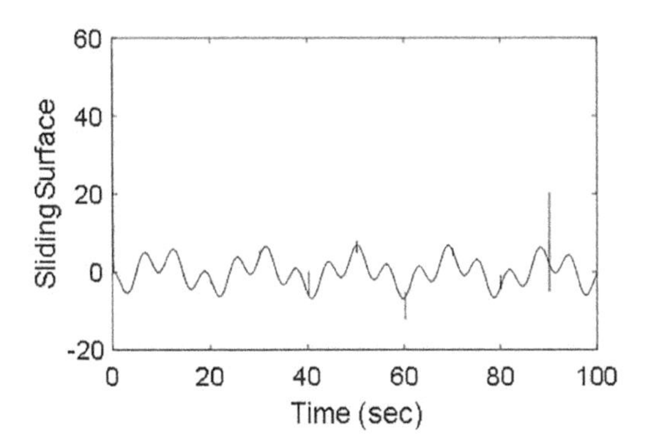

Figure 8.8 Sliding surface (Equation (8.15)).

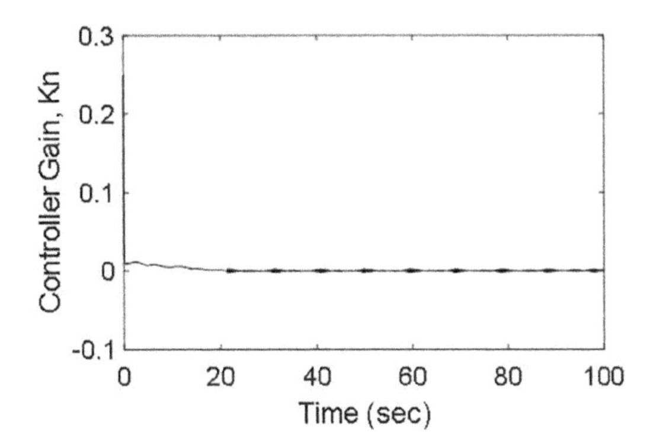

Figure 8.9 Controller gain (Equation (8.27)).

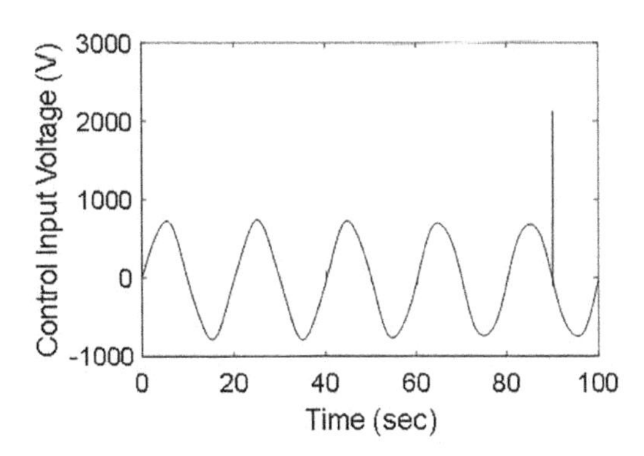

Figure 8.10 Control input voltage (V).

and bounded. These results show that the proposed scheme is an effective approach for robust control application to the HMTS.

8.6 CONCLUSION

This chapter presents the development of a nonlinear robust control scheme for the HMTS. The proposed control scheme has been formulated for HMTS using the BSMC technique. The gain of the proposed controller has been tuned adaptively with the help of the radial basis function neural network. The Lyapunov theory has analyzed the stability of the overall closed-loop HMTS. The simulation study has been carried out for the robust performance analysis of the proposed controller in terms of the degree of tracking, fast convergence of tracking error asymptotically, and bounded control input under system constraints.

REFERENCES

Ding, J., Yang, X., Long, Z., and Dang, N. (2018) "Three-dimensional numerical analysis and optimization of electromagnetic suspension system for 200 km/h maglev train considering eddy current effect," IEEE Access, 6, 61547–61555.

Gilson, F. B. J. and Jose, A. L. B. (2013) "PID control design for a maglev train system," Jornal of Appl. Mech., 389, 425–429.

Hajjaji, A. E., and Ouladsine, M. (2001) "Modeling and nonlinear control of magnetic levitation systems," IEEE Trans. Ind. Electronics., 48(4), 831–838.

Hu, W., Zhou, Y., Zhang, Z. and Fujita, H. (2021), "Model Predictive Control for Hybrid Levitation Systems of Maglev Trains With State Constraints," IEEE Transactions on Vehicular Technology, 70(10), 9972–9985.

Lee, C. Y., Jo, J. M., Han, Y. J., Chung, Y. D., Yoon, Y. S., Choi, S., Hwang, Y. J., Jo, H. C., Jang, J. Y. and Ko, T. K. (2012) "Design, Fabrication, and Operating Test of the Prototype HTS Electromagnet for EMS-Based Maglev," IEEE Transactions on Applied Superconductivity, 22(3), 3600504–3600504.

Lee, C. Y., Jo, J. M., Kang, B., Han, Y. J., Bae, D. K., Yoon, Y. S., Chung, Y. D., Chu, S. Y., Hwang, Y. J., and Ko, T. K. (2011) "Conceptual Design for HTS Coil in Superconducting Electromagnet for Maglev," IEEE Transactions on Applied Superconductivity, 21(3), 1560–1563.

Ni, F., Mu, S., Kang, J., and Xu, J. (2021) "Robust controller design for maglev suspension systems based on improved suspension force model," IEEE

Ni, F., Mu, S., Kang, J., and Xu, J. (2021) "Robust controller design for maglev suspension systems based on improved suspension force model," IEEE Trans.on transportation. Electrification., 7(3), 1765–1779.

Su, X., Yang, X., Shi, P., and Wu, L. (2014) "Fuzzy control of nonlinear electromagnetic suspension systems," Mechatronics, 24(4), 328–335.

Su, X., Yang, X., Shi, P., and Wu, L. (2014), "Fuzzy control of nonlinear electromagnetic suspension systems," Mechatronics, 24(4), 328–335.

Sun, Y., Xu, J., Chen, C., and Lin, G. (2019) "Fuzzy H robust control for magnetic levitation system of maglev vehicles based on TS fuzzy model: Design and experiments," J. Intell. Fuzzy Syst., 36(2), 911–922.

Trans. Transp. Electrific., 7(3), 1765–1779.

Xu, J., Sun, Y., Gao, D., Ma, W., Luo, S., and Qian, Q. (2018) "Dynamic modeling and adaptive sliding mode control for a maglev train system based on a magnetic flux observer," IEEE Access, 6, 31571–31579.

Yang, J., Zolotas, A., Chen, W. H., Michail, K., and Li, S. (2010) "Disturbance observer based control for nonlinear MAGLEV suspension system," Proc. Conf. Control Fault-Tolerant System., Nice, France, 281–286.

Yang, Y. and Ye., Y. (2018), "Backstepping sliding mode control for uncertain strict feedback nonlinear systems using neural-network-based adaptive gain scheduling," Journal of Systems Engineering and Electronics, 29(3), 580–586.

Chapter 9

Reinforcement learning-based intelligent energy management system for electric vehicle

Abhishek Kumar
SOET, CMR University

Amit Kukker
SIT, Symbiosis International

CONTENTS

9.1 INTRODUCTION

Electric vehicles (EVs) have gained a lot of attraction from researchers and industries duc to the limitation of conventional (i.e., fossil) fuels. Moreover, with the increasing use of EVs, the problem of global warming and climate change can be solved to some extent by decreasing the emission of greenhouse gases [1]. EVs can be normally categorized as battery-electric vehicle and hybrid electric vehicles (HEVs) [2]; plug-in hybrid vehicles (PHELs) are an example of HEVs only that includes a rechargeable battery and can be easily connected with external sources. Owing to the advantages of EVs over conventional or internal combustion engine vehicle (ICV) and advancements in power electronics and control strategies [3], most of the automobile industries are moving toward EVs. However, EVs also face some major challenges and one of the most challenging tasks with any EV, HEV, or plug-in hybrid electric vehicle (PHEV) is energy management system (EMS). EVs use rechargeable batteries (to store electrical energy and use this stored energy) along with conventional engine (in the case of HEVs/

DOI: 10.1201/9781003436089-9

PHEVs); hence, an optimization is required to minimize the running cost and fuel used in EVs. Also with advancement in the multigrid system, it is now possible to take energy back from EVs (when not in use) during pick hours and EMS can be used to optimize this energy transfer. A comparative study about various vehicles is given in Table 9.1 [4]. From the table, it is evident that Battery EV (BEV) have the best performance among other vehicles; however, degradation of battery capacity and efficiency with time is a major concern with BEV. To compensate this drawback, hybrid storage system for energy has been studied that consists of battery with supercapacitors [4].

EMS consists of battery management system (BMS) and optimal power flow among various parts of vehicles [3]. In EVs, battery maintenance is very crucial as it accounts for a good percentage of the overall cost of the EV [1] and may cause serious issues related to safety and security. BMS plays a major role as it monitors the battery being used. In the literature, we can find lots of work in the field of EMS for EVs (i.e., all electric vehicles (AEVs), HEVs and PHEVs) [1,5]. Energy management strategies can be used to efficiently utilize stored energy and prevent power losses and battery aging. These can be achieved by considering power splitting, storage characteristics, traffic and driver behavior [4]. In general, EMS can be categorized as (i) rule-based EMS (RBEMS) and (ii) optimized EMS (OEMS) [1]. Intelligent EMS (IEMS) is a category of OEMS [6] that utilizes intelligent optimization and is frequently used [7] as it is adaptive and can adapt to any environment. In general, application of RBEMS, OEMS and IEMS requires a precise model of vehicle, which is very difficult and needs a good amount of cost for parameter calibration. To avoid these issues, model-free control strategy such as reinforcement learning (RL) for EMS can be used [8].

RL is a model-free adaptive learning approach and has been proven very effective in various applications as in medical science [9], control engineering [10,11], power system [12], classification [13,14] and many more. RL is an optimization methodology where reward/punishment strategy is used to choose progressive decisions or actions. This optimization of actions has no set/fixed strategy but it tries to determine the optimal action online via reward/punishment concept. EMS must be designed in such a way that

Table 9.1 Comparative study of various vehicles on road [4]

Technology	*Energy used (kWh/km)*	*Efficiency (In %)*
Petrol	1.36	14
CNG	1.00	19
PHEV	0.42	45
BEV	0.28	67

power and energy competences should be fulfilled for all possible operating conditions including any state of charge (SoC). RL-based EMS, being model free and online learning technique, has shown substantial advantages over conventional and other advanced EMS [15]. EMS architecture basically involves three levels: high, medium and low levels. RL can be implemented on these three levels as per the requirement. As discussed earlier, for RL no prior knowledge of the system is required. Model learn with experience learning and predict future values in a very accurate manner. Similar is applicable for EMS for EV. It will help in prediction of traffic, road conditions and utilization of roads that will cumulatively help in managing energy.

In this review paper, an introduction about EMS and RL is presented in Section 9.1 followed by a brief introduction about EVs and EMS in Section 9.2. Section 9.3 details RL and a detailed review of RL-based EMS is presented in Section 9.4. Finally, we present the conclusion in Section 9.5.

9.2 ELECTRIC VEHICLE AND ENERGY MANAGEMENT SYSTEM

In this section, we shall discuss in detail about EVs along with some important terms used in EVs and then we shall focus on the EMS for EVs.

9.2.1 Electric vehicle

Vehicles can be categorized as ICV, AEV and HEV. While ICV utilizes fuel as a source to run a vehicle, AEV depends only on electric power to run a vehicle. HEVs include both IC engine and electric motor and the extent of hybridization can be measured using the hybridization factor [1]. Hybridization factor can be defined as the ratio of power from the electric motor to the total power that comes from both electric motor and IC engine. Depending on this hybridization factor, HEVs can be categorized as mild HEVs and full HEVs. Full HEVs outperform mild HEVs in terms of fuel use. PHEVs can be considered an extension of full HEVs where battery can be directly connected or plug-in to grid. Various structures of HEVs (i.e., series, parallel, its combination for full HEV and series, parallel for PHEV) can be found in detail in [1], authors in [16] considered series PHEV (as shown in Figure 9.1) for its study.

For any EV/HEV, the following points need attention: (i) units to store energy, for example, battery, supercapacitor and flywheel; (ii) units to generate energy; for example; photovoltaic cell and regenerative braking system; (iii) units to conserve energy; and (iv) technology for charging. In all these points, energy management plays a vital role, for example, SoC, utilization of charging point these all can be efficiently monitored using EMS.

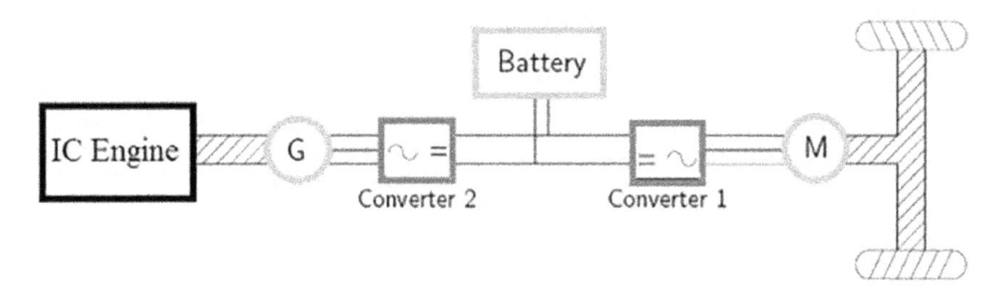

Figure 9.1 Series PHEV.

9.2.2 Energy management system for electric vehicles

To optimize the flow of energy and efficacy of motor and storage system, a fundamental control system is required that can manage all the units and their operations. This need of control unit gives rise to EMS. EMS should be reconfigurable as operating condition keep changing in EVs. In general, EMS has to handle three management tasks [4]:

i. electrical management that considers charging and discharging practices
ii. thermal management to maintain operating temperature and
iii. safety management that looks for the sensors output and takes necessary action in risky situations.

Energy management also refers to power transfer architecture that control energy and power flow. To facilitate reliability of EVs, hybrid energy storage system (HESS) is used that combine more and more units to store energy in the system and bidirectional DC–DC converter to manage energy transfer in two ways [1], i.e., EV to grid and grid to EV. As discussed earlier, this energy management can be rule based, optimization based and AI based as is shown in Figure 9.2. Details about these types can be found in [5–7].

RBEMS mostly relies on experiences of human (gained by their engineering knowledge), even mathematical model, loading strategy on vehicle and driving cycle. In deterministic RBEMS, we have various strategies to handle various conditions [7] as briefed in Table 9.2.

Although implementation of deterministic RBEMS is easy with less computational load but these strategies are not adaptive to change in working environments and don't use real-time data. Fuzzy RBEMS strategies (i) use real-time data, (ii) are robust (i.e., can handle imperfect data) and (iii) adaptive to changing environment. In the literature, we can find many works that imbibe fuzzy logic with EMS. Elkhatib Kamal [17] proposed fuzzy logic rule-based algorithms for hybrid hydraulic-electric vehicles for

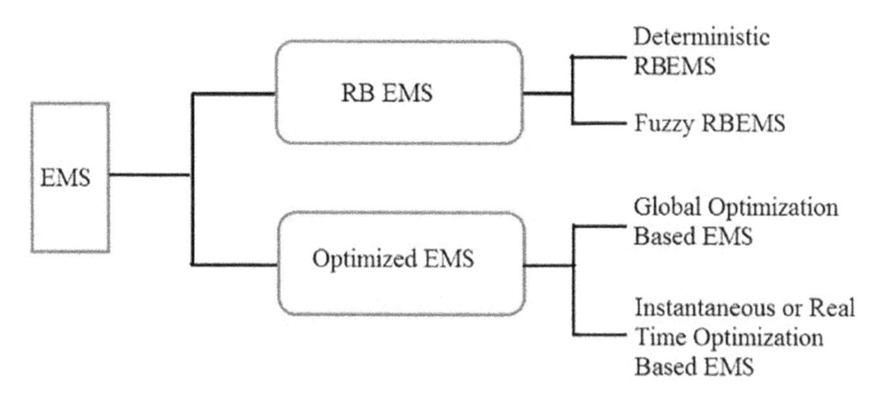

Figure 9.2 Variation of EMS.

Table 9.2 Brief of deterministic RBEMS strategies

Strategy	*Application*	*Limitations*
Thermostat (on/off) strategy	Used on/off strategy depending on pre-set state of charge (SoC) for battery	Low efficiency, ignore power demand to define rule
Power follower	Consider engine as the main source of power, put constraint on power demand by driver.	Doesn't consider fuel consumption and/or emission
Modified power follower	Consider fuel consumption and/or emission apart from other benefits of power follower	Not adaptive
State machine-based strategy	Different approaches that consider driver demand	Optimum utilization of energy is not guaranteed

energy management. In this approach, fuzzy mode switching control is proposed for real-time control and this is possible through fuzzy tuned controllers. This approach provided overall good performance in offline and online control strategy. Fuzzy logic and neural fuzzy logic controller have been proposed in [18] for managing battery SoC. The proposed controller checked accurate torque requirement for energy conversion that charges the battery and decide forward gain for efficient power operation and energy management. In other works, authors have utilized neural network (NN) to introduce robustness, adaptive property and increased efficacy of EMS as in [19] where brake, torque and SoC of battery are considered for formulation of EMS.

OEMS uses the concept of optimizing cost function considering some constraints. The cost function in EMS for EVs may be a combination of SoC, emission, driving performance and some other parameters. In general, the objective function for EMS to be optimized is given as [6]

$$\min J = \int_{t_0}^{t} m(y,u,t)\,dt$$

$$\forall u \in U$$

$$\dot{y} = f(y,u,t) \tag{9.1}$$

$$y(0) = y_0$$

$$y(t) = y$$

where J is the optimization objective function that represents fuel consumption. m, y, u, t_0 and t represent the rate of consumption of fuel, state variable, i.e., SoC of battery, control variable, i.e., output motor torque, initial and final time, respectively. Normally, energy consumption of EV depends on operating temperature, driving profile and SoC.

OEMS can be (i) global optimization-based EMS and (ii) instantaneous or real-time optimization-based EMS. Global optimization-based EMS requires a-priori information of driving cycle [5] and includes optimization methods like dynamic programming (DP) [6], stochastic DP, adaptive fuzzy rule base and linear programming [1]. Guoqiang Li [20] proposed ecological adaptive cruise controller to improve fuel economy. The approach utilized action-dependent heuristic DP that maintained velocity profile for distance control in normal condition. This was an adaptive online approach that deals with disturbances and was designed to control gear shift and power split for fuel consumption. Yonggang Liu [21] proposed cooperative optimization for velocity planning using DP-based EMS strategy. Global optimization-based EMS cannot be used efficiently with real-time scenarios so authors work on instantaneous optimization-based EMS that manages real-time energy to minimum consumption of energy. Instantaneous optimization-based EMS utilizes model predictive control and intelligent control strategies. Teng Liu [22] proposed a genetic algorithm (GA)-assisted controller for online energy management of PHEV-based driving condition. GA was used to find optimal control action offline and was applied to different driving conditions and then activated online. The proposed approach resulted in near global optimization and hardware implementation validated proposed method in real time. The authors [23] have used tri-level game theory for energy management of EV charging station (EVCS) along with EVs as shown in Figure 9.3.

Intelligent EMS can be considered as a part of OEMS only [1,7] as it is optimization using intelligent techniques like particle swarm optimization (PSO), GA, NN and RL. However, some literature [6] assumes intelligent EMS as a different category. IEMS is the advanced EMS strategy that gives adaptive, robust and near to optimum control strategy. Ding [24] proposed rule-based control and GA-based optimization to address battery

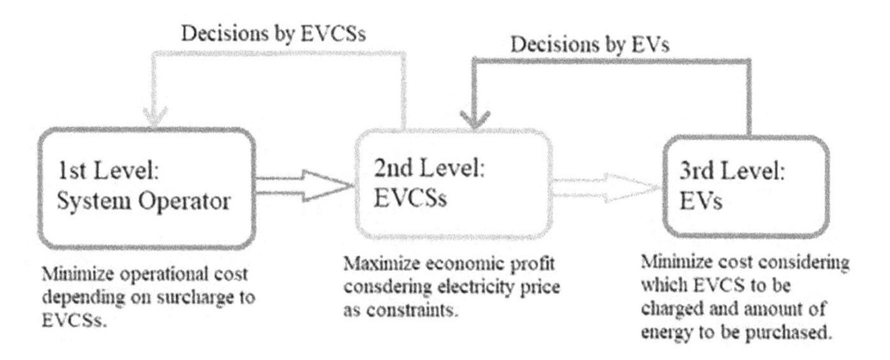

Figure 9.3 Tri-level game theory-based EMS.

limitations of e-vehicles for energy management. Proposed model is tested on ADVISOR database with GA toolbox.

The result shows that GA optimizer successfully achieves its sub-targets in fitness function. Some of the authors used NN to model the driving cycle [6] and other required systems of EVs and then used optimization techniques to design EMS. Table 9.3 briefs some of the EMS techniques used for EVs/PHEVs.

So far we have seen RBEMS and OEMS, RBEMS has the limitation that it is not adaptive and works on predefined rules only. Global OEMS requires prior information of the system to control the EMS, which is not always feasible to get. An online OEMS can work well without the need of prior information about system. However, we get local optima in this case that too on the cost of complex computation and high dependency on the unidentified trip information. RL-based EMS strategy provides a trade-off between computation time, real-time implementation and optimality of solution [29]. So most of the work, nowadays, depends on RL-based EMS strategy for all kinds of EVs.

9.3 REINFORCEMENT LEARNING BASICS

RL is a paradigm of machine learning and works on the principle of learning by interaction. In brief, RL has an agent (that can be considered as decision maker) and this agent interacts with environment that it wants to control. In this interaction, a reward is assigned to agent depending on the decision taken by agent. This reward or reward function is framed to be the cost function of optimization problem and agent's decision sequence (also known as policy) is to optimize this reward in the long run [30]. Thus, we can say that in RL, agent select action (a_t) when the environment is in state (S_t) and this action results in (i) the change of environmental state from S_t to S_{t+1} and (ii) a reward function r_{t+1}. This reward function is basically

Table 9.3 Brief of various EMS techniques

References	*Technique*	*Remarks*
Azidin [25]	Rule-based control strategy for EMS	Focused on light hybrid EV where acceleration and load power were optimized by switching between various energy resources (fuel cell, batter and supercapacitor).
Elkhatib Kamal [17]	Fuzzy logic basic switching	Adaptive, online strategy that doesn't require prior knowledge. Minimize aging effect and operated near optimal range.
Krishna Veer Singh [19]	Fuzzy logic and Elman NN model	Provided less SoC degradation and increased fuel economy in real time.
Guoqiang Li [20]	Heuristic dynamic programming	Adaptive online approach that can handle disturbances also. Designed in a manner to control gear shift and power split for fuel consumption.
Teng Liu [22]	Genetic algorithm-assisted controller	Hardware implementation validated the proposed method in real-time characteristic of strategy.
Ding [24]	Rule-based control and GA-based optimization	Improvement in results in the manner of HC and NOx emission.
Dapai Shi [6]	NN and GA	Used NN to construct driving cycle recognizer and GA as optimizer. Fuel economy was improved by 3%.
Florence Berthold [16], Hamid Khayyam [26]	Fuzzy logic-based controller	Optimized SoC for Vehicle to Home (V2H) [16] and Vehicle to Grid (V2G) [26] for PHEV.
Mayank Jha [27]	IEMS using adaptive PSO	Controlled bidirectional flow of power between distributed generators and load for V2G. Resulted in less switching time.
Salvatti [28]	Dynamic programming	Optimized charging and discharging profile resulting in improved efficiency of microgrid.
Bahram Shakerighadi [23]	Game theory-based EMS	Energy management of EVCS along with EVs to optimize profit (in terms of cost) and efficient utilization of energy.

an objective function in RL and RL agent tries to optimize it by selecting an appropriate sequence of action. Figure 9.4 shows RL-based EMS that considered engine power, speed and SoC as state of the environment and engine power as decision or action [2]. Here, EMS acts as an agent and selects engine power such that fuel consumption is reduced without over charging and discharging of battery and maintaining SoC of battery in the safe range. The authors [2] used the reward function as

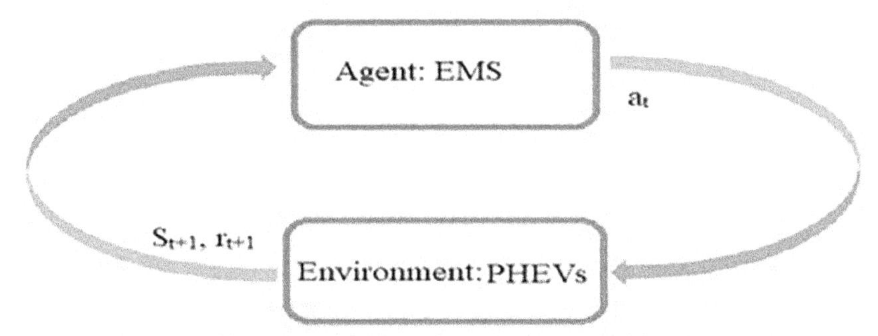

Figure 9.4 Reinforcement learning-based EMS.

$$R = -\int_{t_0}^{t} \rho\, m(t)\, dt + \sigma\left[soc(t) - soc(t_0)\right]^2 \tag{9.2}$$

The reward function, defined here, consists of fuel economy/efficiency/consumption (rate of consumption of fuel) described by $m(t)$ and SoC performance denoted by SoC difference between present time t and initial time t_0. The two parameters ρ and σ can be adjusted to trade-off SoC and fuel economy of the EV. SoC can be defined as a fraction of available battery power to full power of the battery and can be given by [6]

$$\dot{SoC} = \begin{cases} -\dfrac{V_{OC} - \sqrt{\left(V_{OC}^2 - 4P_bR_0\right)}}{R_0Q_b}\eta_b, & \text{discharge} \\[2ex] \dfrac{V_{OC} - \sqrt{\left(V_{OC}^2 - 4P_bR_0\right)}}{R_0Q_b\eta_b}, & \text{charge} \end{cases} \tag{9.3}$$

Here, η_b, Q_b, R_0, P_b and V_0 represent charging–discharging efficiency, total capacity of battery power, internal resistance, available power of battery and open circuit voltage.

RL can be (i) model based where an agent (EMS in the case of EVs) optimizes the performance of EVs by estimating the model of environment (EVs and its environment) as Markov Decision Process (MDP) or (ii) model free where agent learns optimal policy online while interacting with the environment. RL also has many variants like Q-learning, SARSA, TD(λ) and more. The authors have applied variations of RL in EMS problem and will be discussed in detail in the next section.

9.4 REINFORCEMENT LEARNING-BASED ENERGY MANAGEMENT SYSTEM FOR ELECTRIC VEHICLE

In RL-based EMS, EMS can be modeled as MDP where decision maker, i.e., energy management strategy, can be considered an agent and the EVs/HEVs/PHEVs can be considered environment [2]. There may be variations in the structure of RL-based EMS as per the requirement. Guodong Du [31] proposed an online energy management approach to reduce fuel consumption of hybrid electric tracked vehicles via fast Q-learning. In this, fast Q-learning-based online updating of control policy has reduced the consumption by 4.6% and computational time is 1.35 sec which will be approximately 16% higher than the DP. The result is tested on three real-time driving schedules. This can be implemented in real time in near future after validation on more real-time scenarios. Liu [8] proposed a heuristic planning controller for energy management of PHEV. The approach is model-free online Q-learning having Dyna agent name as Dyna-H algorithm. This approach is validated on all control components of a Chevrolet Volt where four traction operation modes are applied. The testing result shows that the algorithm has definitely improved fuel consumption and significantly improved computational speed as compared to Q-learning and Dyna-algorithm. A transfer learning technique using deep reinforcement learning (DRL) [32] was suggested to get energy management strategies that can boost the performance of EMS for HEV. A hybrid battery system for energy management in EV via DRL has been used in [15] for high-energy and high-power battery ratings. This approach is based on battery cells electrical and thermal characterization to minimize energy loss and improve electrical and thermal safety levels. This approach shows superiority over the RL—in computation time and energy loss reduction for hybrid battery systems. Haochen Sun [33] proposed fuel cell energy management of EVs equipped with three different power sources. Equivalent Consumption Minimization Strategy (ECMS) is proposed to find the trade-off between global and real-time design through RL. Power splitting policy provides different driving cycles and traffic conditions to achieve energy management. The result shows high computation efficiency, lower power fluctuation and optimal fuel economy.

Yuecheng Li [34] proposed adaptive EMSs for hybrid vehicles using Deep RL. This method utilizes the previous history data to take future decision-based driving information so that the generalized energy management solution is formulated with the help of MDP. The method introduced offline training to online application because of its learning and optimal decision capability. When tested on benchmark problem shows 3.5% better result than the existing MPS approach with an impressive computation speed of .001 sec showing its practical potential. Also, future driving is possible even when driving information is not available. Yuankai Wu [35] proposed a continuous control strategy of energy management for plug-in hybrid electric bus based on traffic information using DRL. Based on different traffic simulation for driving cycles, optimized model is derived that also have the information about the on-board passengers. The main contribution was to explore and incorporate traffic information with HEV via advanced intelligent algorithms for energy management. Bin Xu [36] gives parametric study-based optimized energy management for hybrid e-vehicle. Bin Xu [37] proposed an ensemble RL strategy for hybrid vehicle fuel economy based on supervisory control. In this, multiple agent with different state combinations are used. Two energy management strategies named as thermostatic and equivalent consumption minimization are used. The action is taken jointly by weighted average of these strategies. With this approach, fuel economy is 3.2% higher than the best of single agent. In [38], DRL is used to autonomously learn optimal fuel splits with the interaction between traffic environment and vehicle. This model proves to save 16.3% fuel compared to conventional binary control strategies. Chang Liu [39] proposed Q-learning and neuro DP-based optimal energy management for HEVs. The approach helped to estimate the expected future energy cost of vehicle based on state and control actions. A new initialization strategy was also introduced which reduced learning convergence by 70% and provide optimal learning with proper selection of penalty function. Hongwen He [40] proposed a pedal control strategy to optimize energy consumption of EVs acceleration process. Based on the training of Q-learning relation between vehicle acceleration and proportion of energy consumption reduction is analyzed with respect to the time. To improve pedal control, DQN is proposed. The overall approach not only achieved energy saving but also stability of control.

The authors have also used a double layer intelligent energy management approach [41] for optimal PEV integration. Xuewei Qi [42] proposed a DRL approach that learns autonomously from previous driving records and takes optimal fuel usage actions. This method is fully data-driven, i.e., no predefined rules or prediction is required. The approach achieved 16.3% fuel economy improvement as compared to binary control strategies. Yuan Zou [43] proposed a systematic control-oriented model for optimal energy utilization and used Kullback–Leibler divergence rate in real-time application. RL is applied to find an optimized control strategy resulting in significant improvement in fuel economy and efficiency in real time. In another approach [44], fuzzy encoding and nearest neighbor approach are used for velocity prediction. For optimal control between power distribution and demand, RL management is used. This approach has reduced significantly fuel consumption and computation time compared to DP. Jingda Wu [45] proposed a DRL approach control that improved the performance of EV without the need to clarify about complex internal factors in controlled objects. Chao Yang [46] proposed an adaptive firework algorithm for control strategy where T-S fuzzy control-based torque distribution is used to optimize engine operating points as per the required torque of powertrain and battery SOC. The method is validated through simulation and HIL test.

Guodong Du [47] suggested double DQL for powertrain model for series HEV. Modified prioritized experience replay and adaptive optimizer method is used to update network weights called AMSGrad. The proposed method attains better training efficiency and lower energy consumption that were nearly equal to global optimal solution. Fuel consumption reduction via indirect RL technique is proposed in [48]. High order Markov chain model is used for power transition probability value updating online and recursively for control policy. Chunyang Qi [49] proposed a self-supervised RL method. This technique is based on DRL for fuel consumption economy of PHEV. In this, powertrain model is designed and enriched through reward function that is given by self-supervised learning. The result shows fast training convergence and less fuel consumption compared to existing traditional methods which are approximately near to global optimal solution to new driving cycles. Q-Dyna RL [50] algorithm for EMS reduced learning time and improved control performance. The result shows fast learning and maintains fuel consumption in real-time performance. A brief of some selected work related to the application of RL in intelligent EMS is listed in Table 9.4.

Table 9.4 Brief about use of RL in intelligent EMS for EVs and its variants

Author	*Method/technique used*	*Remarks/result*
Guodong Du [31]	Fast Q-learning	Fuel consumption reduced by 4.6% and computational time is 1.35 sec.
Teng Liu [8]	Q-learning having Dyna agent name as Dyna-H algorithm	Validated on all control components of a Chevrolet, improved fuel consumption and significantly improved computational speed.
Riadh Abdelhedi [29]	RL-based instantaneous EMS	Reduced RMS battery current under varying load for HESS consisting of battery and supercapacitor.
Renzong Lian [32]	Transfer learning techniques using DRL	Improved convergence efficiency.
Weihan Li [15]	DRL	Improved thermal and electrical safety for hybrid battery systems.
Xuewei Qi [51]	Real-time energy management using RL	Around 12% fuel saving is achieved in with charging consideration and additional 8% saving when charging is compared to standard binary mode strategy.

(Continued)

Table 9.4 (*Continued*) Brief about use of RL in intelligent EMS for EVs and its variants

Author	*Method/technique used*	*Remarks/result*
Yuecheng Li [34]	DRL	Adaptive EMS tested on benchmark problem showing a remarkable computation speed of .001 sec.
Bin Xu [37]	Ensemble RL	Fuel economy efficacy obtained was 3.2% higher than the best of single agent.
Xuewei Qi [38]	Self-learning control for energy efficient driving using DRL	This model saved 16.3% fuel compared to conventional control strategies. Also, a dueling DQN is implemented that gave better convergence as compared to single DQN.
Chang Liu [39]	Q-learning and neuro dynamic programming	Based on fixed and random driving cycles. An initialization strategy is introduced to give optimal learning with proper selection of penalty function.
Xuewei Qi [52]	Evolutionary algorithms	Self-adaptive SOC is best for real-time traffic situation and robust in manner of uncertainties in recharging possibilities. Gave very good fuel economy, especially with minimum driving knowledge.
Jingda Wu [53]	Deep deterministic policy gradient approach in combination of expert-assistance system.	Improved thermal safety and reduced driving cost.
Xinyou Lin [54]	Improved RL	Exploration factor (EF) is considered here to address RL convergence and reward cost evaluation. KL divergence provides optimal solutions. Improved RL and EF shows significant improvement in the efficiency of PHEV.
Jianhao Zhou [55]	DRL algorithm TD3	Eliminate irrational torque allocation using heuristic rule-based controller, environment disturbance mitigated by hybrid experience.
Guodong Du [56]	Heuristic DRL. For better optimization. Nesterov accelerated gradient method is used.	High precision driving cycle is possible. Method adaptability, stability and robustness are tested.

9.5 CONCLUSION

A survey of intelligent EMS is presented in this chapter with special emphasis on the use of RL. We focused on the use and types of EVs along with some challenges faced by EVs for its efficient use in practical life. We find the importance of EMS in the EVs/HEVs/PHEVs. The survey suggested the use of intelligent EMS to get an almost optimal solution for energy management under changing environment. We focused mainly on RL and its variant to design EMS resulting in efficient fuel economy, better SoC, improved thermal safety and fast computation. We also observed improvement and progress in V2H or V2G technology with better strategies for charging and discharging. Optimum use of EMS for EVs and its variant can be obtained with proper modeling of uncertainties associated with environment and traffic/driving pattern. In future, more autonomy and better prediction algorithm can be used in energy management strategies for EVs especially in PHEVs.

APPENDIX-I

EV	Electric vehicle
EMS	Energy management system
RL	Reinforcement learning
PHEV	Plug-in hybrid electric vehicle
ICV	Internal combustion engine vehicle
HEV	Hybrid electric vehicle
BMS	Battery management system
HESS	Hybrid energy storage system
RBEMS	Rule-based EMS
OEMS	Optimized EMS
SoC	State of charge
PSO	Particle swarm optimization
GA	Genetic algorithm
NN	Neural network
DQN	Deep Q-Learning
DRL	Deep reinforcement learning

REFERENCES

1. Tie, Siang Fui, and Chee Wei Tan. "A review of energy sources and energy management system in electric vehicles." *Renewable and sustainable energy reviews* 20(2013): 82–102.
2. Qi, Chunyang, Yiwen Zhu, Chuanxue Song, Guangfu Yan, Feng Xiao, Xu Zhang, Jingwei Cao, and Shixin Song. "Hierarchical reinforcement learning based energy management strategy for hybrid electric vehicle." *Energy* 238 (2022): 121703.
3. Aruna, P., and Prabhu V. Vasan. "Review on energy management system of electric vehicles." In *2019 2nd International Conference on Power and Embedded Drive Control (ICPEDC)*, pp. 371–374. IEEE, 2019.
4. Rimpas, Dimitrios, Stavros D. Kaminaris, Izzat Aldarraji, Dimitrios Piromalis, Georgios Vokas, Panagiotis G. Papageorgas, and Georgios Tsaramirsis. "Energy management and storage systems on electric vehicles: A comprehensive review." *Materials Today: Proceedings* (2021).

5. Yang, Chao, Mingjun Zha, Weida Wang, Kaijia Liu, and Changle Xiang. "Efficient energy management strategy for hybrid electric vehicles/plug-in hybrid electric vehicles: review and recent advances under intelligent transportation system." *IET Intelligent Transport Systems* 14, no. 7 (2020): 702–711.
6. Shi, Dapai, Shipeng Li, Kangjie Liu, Yinggang Xu, Yun Wang, and Changzheng Guo. "Adaptive energy management strategy for plug-in hybrid electric vehicles based on intelligent recognition of driving cycle." *Energy Exploration & Exploitation* (2022): 01445987221111488.
7. Ostadian, Reihaneh, John Ramoul, Atriya Biswas, and Ali Emadi. "Intelligent energy management systems for electrified vehicles: Current status, challenges, and emerging trends." *IEEE Open Journal of Vehicular Technology* 1(2020): 279–295.
8. Liu, Teng, Xiaosong Hu, Weihao Hu, and Yuan Zou. "A heuristic planning reinforcement learning-based energy management for power-split plug-in hybrid electric vehicles." *IEEE Transactions on Industrial Informatics* 15, no. 12 (2019): 6436–6445.
9. Coronato, Antonio, Muddasar Naeem, Giuseppe De Pietro, and Giovanni Paragliola. "Reinforcement learning for intelligent healthcare applications: A survey." *Artificial Intelligence in Medicine* 109 (2020): 101964.
10. Kumar, Abhishek, and Rajneesh Sharma. "Neural/fuzzy self learning Lyapunov control for non linear systems." *International Journal of Information Technology* (2018): 1–14.
11. Kumar, Abhishek, and Rajneesh Sharma. "A stable Lyapunov constrained reinforcement learning based neural controller for non linear systems." In *International conference on computing, communication & automation*, (2015): pp. 185–189.
12. Navin, Nandan Kumar, and Rajneesh Sharma. "A fuzzy reinforcement learning approach to thermal unit commitment problem." *Neural Computing and Applications* 31, no. 3 (2019): 737–750.
13. Kukker, Amit, Rajneesh Sharma, and Hasmat Malik. "An intelligent genetic fuzzy classifier for transformer faults." *IETE Journal of Research* (2020): 1–12.
14. Kukker, A. and Rajneesh Sharma, "JAYA-Optimized Fuzzy Reinforcement Learning Classifier for COVID-19." *IETE Journal of Research* (2022): pp.1–12.
15. Li, Weihan, Han Cui, Thomas Nemeth, Jonathan Jansen, Cem Uenluebayir, Zhongbao Wei, Lei Zhang et al. "Deep reinforcement learning-based energy management of hybrid battery systems in electric vehicles." *Journal of Energy Storage* 36(2021): 102355.
16. Berthold, Florence, Alexandre Ravey, Benjamin Blunier, David Bouquain, Sheldon Williamson, and Abdellatif Miraoui. "Design and development of a smart control strategy for plug-in hybrid vehicles including vehicle-to-home functionality." *IEEE Transactions on Transportation Electrification* 1, no. 2 (2015): 168–177.
17. Kamal, Elkhatib, Lounis Adouane, Rustem Abdrakhmanov, and Nadir Ouddah. "Hierarchical and adaptive neuro-fuzzy control for intelligent energy management in hybrid electric vehicles." IFAC-*PapersOnLine* 50, no. 1 (2017): 3014–3021.

18. Suhail, Mohammad, Iram Akhtar, Sheeraz Kirmani, and Mohammed Jameel. "Development of progressive fuzzy logic and ANFIS control for energy management of plug-in hybrid electric vehicle." *IEEE Access* 9(2021): 62219–62231.
19. Singh, Krishna Veer, Hari Om Bansal, and Dheerendra Singh. "Fuzzy logic and Elman neural network tuned energy management strategies for a power-split HEVs." *Energy* 225(2021): 120152.
20. Li, Guoqiang, and Daniel Görges. "Ecological adaptive cruise control and energy management strategy for hybrid electric vehicles based on heuristic dynamic programming." IEEE Transactions on Intelligent Transportation Systems 20, no. 9 (2018): 3526–3535.
21. Liu, Yonggang, Zhenzhen Huang, Jie Li, Ming Ye, Yuanjian Zhang, and Zheng Chen. "Cooperative optimization of velocity planning and energy management for connected plug-in hybrid electric vehicles." *Applied Mathematical Modelling* 95(2021): 715–733
22. Liu, Teng, Huilong Yu, Hongyan Guo, Yechen Qin, and Yuan Zou. "Online energy management for multimode plug-in hybrid electric vehicles." *IEEE Transactions on Industrial Informatics* 15, no. 7 (2018): 4352–4361.
23. Shakerighadi, Bahram, Amjad Anvari-Moghaddam, Esmaeil Ebrahimzadeh, Frede Blaabjerg, and Claus Leth Bak. "A hierarchical game theoretical approach for energy management of electric vehicles and charging stations in smart grids." *IEEE Access* 6 (2018): 67223–67234.
24. Ding, N., K. Prasad, and T. T. Lie. "Design of a hybrid energy management system using designed rule-based control strategy and genetic algorithm for the series-parallel plug-in hybrid electric vehicle." *International Journal of Energy Research* 45, no. 2 (2021): 1627–1644.
25. Azidin, F. A., Z. A. Ghani, M. A. Hannan, and M. Azah. "Intelligent Control Algorithm for Energy Management System of Light Electric Vehicles." *Journal of Telecommunication, Electronic and Computer Engineering (JTEC)* 6, no. 2 (2014): 37–43.
26. Khayyam, Hamid, Hassan Ranjbarzadeh, and Vincenzo Marano. "Intelligent control of vehicle to grid power." *Journal of Power Sources* 201(2012): 1–9.
27. Jha, Mayank, Frede Blaabjerg, Mohammed Ali Khan, Varaha Satya Bharath Kurukuru, and Ahteshamul Haque. "Intelligent control of converter for electric vehicles charging station." *Energies* 12, no. 12 (2019): 2334.
28. Salvatti, Gabriel Antonio, Emerson Giovani Carati, Rafael Cardoso, Jean Patric da Costa, and Carlos Marcelo de Oliveira Stein. "Electric vehicles energy management with V2G/G2V multifactor optimization of smart grids." *Energies* 13, no. 5 (2020): 1191.
29. Abdelhedi, Riadh, Amine Lahyani, Ahmed Chiheb Ammari, Ali Sari, and Pascal Venet. "Reinforcement learning-based power sharing between batteries and supercapacitors in electric vehicles." In *2018 IEEE International Conference on Industrial Technology (ICIT)*, (2018), pp. 2072–2077.
30. Kumar, A. and R. Sharma, "Linguistic Lyapunov reinforcement learning control for robotic manipulators." *Neurocomputing*, 272, (2018): 84–95.
31. Du, Guodong, Yuan Zou, Xudong Zhang, Zehui Kong, Jinlong Wu, and Dingbo He. "Intelligent energy management for hybrid electric tracked vehicles using online reinforcement learning." *Applied Energy* 251(2019): 113388.

32. Lian, Renzong, Huachun Tan, Jiankun Peng, Qin Li, and Yuankai Wu. "Cross-type transfer for deep reinforcement learning based hybrid electric vehicle energy management." *IEEE Transactions on Vehicular Technology* 69, no. 8 (2020): 8367–8380.
33. Sun, Haochen, Zhumu Fu, Fazhan Tao, Longlong Zhu, and Pengju Si. "Data-driven reinforcement-learning-based hierarchical energy management strategy for fuel cell/battery/ultracapacitor hybrid electric vehicles." *Journal of Power Sources* 455(2020): 227964.
34. Li, Yuecheng, Hongwen He, Jiankun Peng, and Hong Wang. "Deep reinforcement learning-based energy management for a series hybrid electric vehicle enabled by history cumulative trip information." *IEEE Transactions on Vehicular Technology* 68, no. 8 (2019): 7416–7430.
35. Wu, Yuankai, Huachun Tan, Jiankun Peng, Hailong Zhang, and Hongwen He. "Deep reinforcement learning of energy management with continuous control strategy and traffic information for a series-parallel plug-in hybrid electric bus." *Applied energy* 247(2019): 454–466.
36. Xu, Bin, Dhruvang Rathod, Darui Zhang, Adamu Yebi, Xueyu Zhang, Xiaoya Li, and Zoran Filipi. "Parametric study on reinforcement learning optimized energy management strategy for a hybrid electric vehicle." *Applied Energy* 259(2020): 114200.
37. Xu, Bin, Xiaosong Hu, Xiaolin Tang, Xianke Lin, Huayi Li, Dhruvang Rathod, and Zoran Filipi. "Ensemble reinforcement learning-based supervisory control of hybrid electric vehicle for fuel economy improvement." *IEEE Transactions on Transportation Electrification* 6, no. 2 (2020): 717–727.
38. Qi, Xuewei, Yadan Luo, Guoyuan Wu, Kanok Boriboonsomsin, and Matthew Barth. "Deep reinforcement learning enabled self-learning control for energy efficient driving." *Transportation Research Part C: Emerging Technologies* 99(2019): 67–81.
39. Liu, Chang, and Yi Lu Murphey. "Optimal power management based on Q-learning and neuro-dynamic programming for plug-in hybrid electric vehicles." *IEEE transactions on neural networks and learning systems* 31, no. 6 (2019): 1942–1954.
40. He, Hongwen, Jianfei Cao, and Xing Cui. "Energy optimization of electric vehicle's acceleration process based on reinforcement learning." *Journal of Cleaner Production* 248(2020): 119302.
41. Mehta, Rahul, Pranjal Verma, Dipti Srinivasan, and Jing Yang. "Double-layered intelligent energy management for optimal integration of plug-in electric vehicles into distribution systems." *Applied energy* 233(2019): 146–155.
42. Qi, Xuewei, Yadan Luo, Guoyuan Wu, Kanok Boriboonsomsin, and Matthew J. Barth. "Deep reinforcement learning-based vehicle energy efficiency autonomous learning system." In *2017 IEEE intelligent vehicles symposium* (IV), pp. 1228–1233. IEEE, 2017.
43. Zou, Yuan, Teng Liu, Dexing Liu, and Fengchun Sun. "Reinforcement learning-based real-time energy management for a hybrid tracked vehicle." *Applied energy* 171(2016): 372–382.
44. Liu, Teng, Xiaosong Hu, Shengbo Eben Li, and Dongpu Cao. "Reinforcement learning optimized look-ahead energy management of a parallel hybrid electric vehicle." *IEEE/ASME Transactions on Mechatronics* 22, no. 4 (2017): 1497–1507.

45. Wu, Jingda, Hongwen He, Jiankun Peng, Yuecheng Li, and Zhanjiang Li. "Continuous reinforcement learning of energy management with deep Q network for a power split hybrid electric bus." *Applied energy* 222(2018): 799–811.
46. Yang, Chao, Kaijia Liu, Xiaohong Jiao, Weida Wang, Ruihu Chen, and Sixiong You. "An adaptive firework algorithm optimization-based intelligent energy management strategy for plug-in hybrid electric vehicles." *Energy* 239(2022): 122120.
47. Du, Guodong, Yuan Zou, Xudong Zhang, Lingxiong Guo, and Ningyuan Guo. "Energy management for a hybrid electric vehicle based on prioritized deep reinforcement learning framework." *Energy* 241(2022): 122523.
48. Yang, Ningkang, Lijin Han, Changle Xiang, Hui Liu, and Xunmin Li. "An indirect reinforcement learning based real-time energy management strategy via high-order Markov Chain model for a hybrid electric vehicle." *Energy* 236(2021): 121337.
49. Qi, Chunyang, Yiwen Zhu, Chuanxue Song, Jingwei Cao, Feng Xiao, Xu Zhang, Zhihao Xu, and Shixin Song. "Self-supervised reinforcement learning-based energy management for a hybrid electric vehicle." *Journal of Power Sources* 514(2021): 230584.
50. Yang, Ningkang, Lijin Han, Changle Xiang, Hui Liu, and Xuzhao Hou. "Energy management for a hybrid electric vehicle based on blended reinforcement learning with backward focusing and prioritized sweeping." *IEEE Transactions on Vehicular Technology* 70, no. 4 (2021): 3136–3148.
51. Qi, Xuewei, Guoyuan Wu, Kanok Boriboonsomsin, Matthew J. Barth, and Jeffrey Gonder. "Data-driven reinforcement learning–based real-time energy management system for plug-in hybrid electric vehicles." *Transportation Research Record* 2572, no. 1 (2016): 1–8.
52. Qi, Xuewei, Guoyuan Wu, Kanok Boriboonsomsin, and Matthew J. Barth. "Development and evaluation of an evolutionary algorithm-based online energy management system for plug-in hybrid electric vehicles." *IEEE Transactions on Intelligent Transportation Systems* 18, no. 8 (2016): 2181–2191.
53. Wu, Jingda, Zhongbao Wei, Kailong Liu, Zhongyi Quan, and Yunwei Li. "Battery-involved energy management for hybrid electric bus based on expert-assistance deep deterministic policy gradient algorithm." *IEEE Transactions on Vehicular Technology* 69, no. 11 (2020): 12786–12796.
54. Lin, Xinyou, Kuncheng Zhou, Liping Mo, and Hailin Li. "Intelligent energy management strategy based on an improved reinforcement learning algorithm with exploration factor for a plug-in PHEV." *IEEE Transactions on Intelligent Transportation Systems* (2021).
55. Zhou, Jianhao, Siwu Xue, Yuan Xue, Yuhui Liao, Jun Liu, and Wanzhong Zhao. "A novel energy management strategy of hybrid electric vehicle via an improved TD3 deep reinforcement learning." *Energy* 224(2021): 120118.
56. Du, Guodong, Yuan Zou, Xudong Zhang, Lingxiong Guo, and Ningyuan Guo. "Heuristic energy management strategy of hybrid electric vehicle based on deep reinforcement learning with accelerated gradient optimization." *IEEE Transactions on Transportation Electrification* 7, no. 4 (2021): 2194–2208.

Chapter 10

Soft-computing-based discharge parameters estimator for Li-ion batteries

Shubham Kumar, Shivam Sharma, Tarun Kumar, Brijesh Singh, and Anmol Gupta
KIET Group of Institutions

CONTENTS

10.1 INTRODUCTION

With rising costs, depletion of natural resources, carbon emissions [1], and the availability of conventional fuels, it is necessary to find more efficient, clean, safe, and economical energy sources and arrange for their subsequent storage. As a result, battery technology was started and developed, and much work has been done in this field. In 1800, Alessandro Volta invented the first electrochemical battery [2]. It was a stack of copper and zinc plates separated by salty water-soaked paper discs that could keep a continuous current stream going for a long period. Although early batteries were crucial for experiments, their voltages fluctuated, and they could not produce a big current for long periods [3]. The Daniel cell, created in 1836 by British chemist John Frederick Daniels, was the first practical electricity generator widely embraced and acknowledged as an industry standard. [4].

Batteries are one of the oldest technologies in the realm of energy storage. A battery is made up of many electrochemical cells. A battery can contain an infinite number of cells. The capacity of a battery is measured in ampere-hours (Ah). Primary and secondary batteries are the two types of batteries available. Primary batteries can only be charged once and are disposed of when their capacity is completely depleted.

On the other hand, secondary batteries can be recharged and reused several times. The battery cycle refers to the number of charging and

DOI: 10.1201/9781003436089-10

discharging cycles. Annual battery consumption climbed by 30% between 2010 and 2018, reaching 180 GWh in 2018. The growth rate will be kept at a conservative 25%, with demand reaching 2600 GWh in 2030. Furthermore, cost reductions are predicted to boost demand to 3562 GWh [5]. The electrification of transportation, [5] the large-scale deployment of electricity grids [5], and the transition away from fossil fuel combustible energy sources encouraged by anthropogenic climate change are all important causes for the electric battery industry's rapid expansion. Clean, renewable energy sources, and stricter emissions regulations are the way to go. The linear battery model, often known as the internal resistance (IR) model [6], is a simple battery system model containing a resistance Rint and a voltage source, Voc (Figure 10.1).

The resistance Rint represents the energy losses, which make batteries heat up. Terminal voltage VT matches up with open-circuit voltage VOC only in an open circuit. However, when a load is connected, the voltage is given by [7]

$$V_T = V_{OC} - R_{\text{int}} * I \tag{1}$$

The present study aims to find out the change in discharge parameters of Li-ion batteries. For this study, the help of MATLAB® simulation has been taken. Lithium-ion batteries have two types of parameters: normal parameters and discharge parameters [8]. A variety of common parameters are nominal voltage, rated capacity, response time, and state of charge [9]. Discharge parameters are those which are responsible for discharging the battery. The various discharge parameters are cutting voltage [9], fully charged voltage [10], nominal discharge current [9], capacity at nominal

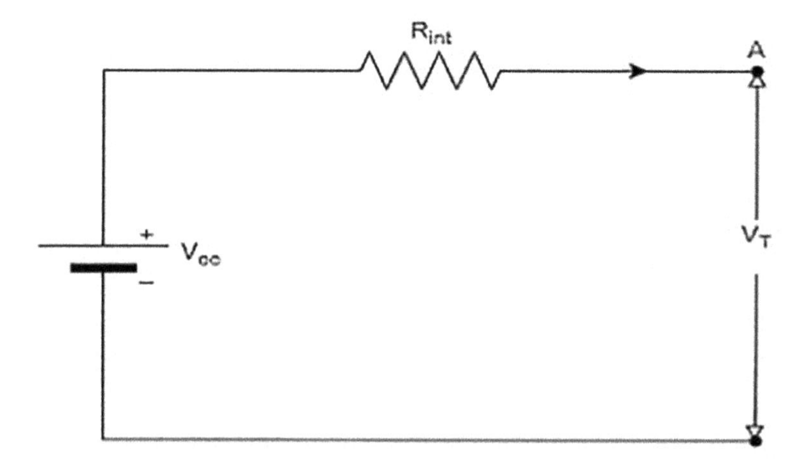

Figure 10.1 Model of a linear battery.

voltage Ah [8], exponential zone [8], and IR [11]. It is suggested that the performance of batteries in different applications can be changed by changing or controlling the discharge parameters [12]. Simulation models of different real Li-ion batteries have been prepared, and their discharge parameters have been studied. The simulation results show that the discharge parameter of all Li-ion batteries deviates from a certain percentage value. This estimation method can also be used for battery selection.

10.2 PERFORMANCE OF LI-ION BATTERIES: TECHNICAL REVIEW

For better performance, selecting the right battery is essential. There are many different types of batteries on the market. Every battery has some level of quality. The most common battery types are Ni-M-H (nickel-metal hydride), lithium-ion, ultracapacitor, and lead-acid batteries. These are the four most often used batteries, and the table below compares them. The Li-ion batteries have an advantage due to good energy efficiency, better temperature operating range, lower weight, and better cycle life. Based on a literature survey and comparison, it has been found that lithium-ion batteries are better than other batteries for further research [13–19]. Considering the quality of Li-ion batteries, locating the right battery based on discharge parameters is a difficult practical process; it can be obtained only in advanced labs. At the same time, it is usually difficult for consumers to select the right batteries or estimate the discharge parameters. Based on the literature survey and comparisons in Table 10.1. It has been discovered that lithium-ion batteries outperform other batteries. Lithium-ion batteries have the following benefits:

- long cycle life,
- high specific energy and energy density,
- broad temperature range of operation,
- no memory effect,
- long shelf life,
- low local action rate,
- low self-discharge rate,

Table 10.1 Comparison of various batteries

Battery	*Easy access/ Inexpensive*	*Energy efficient*	*Temperature performance*	*Weight*	*Life cycle*
Lithium-ion	✓	✓	✓	✓	✓
Nickel-Metal Lead-Acid	✗	✓	✗	✓	✗
Ultracapacitor	✓	✓	✗	✓	✓
	✗	✓	✓	✓	✗

- high voltage per monomer,
- rapid charge capability, high energy to volume,
- high rate and high power discharge,
- high energy to quality capability.

Lithium batteries are frequently used in wireless base stations as a backup power supply and power source in electric cars. Along with its benefits, the Li-ion battery has certain drawbacks, including the inability to resist high currents (several times its maximum current capacity). Lithium batteries' anisotropic thermal characteristic [20] is also a concern. Furthermore, overcharging or discharging a battery raises its internal temperature [21]. It cannot resist high temperatures and explodes when abused [22]. The more batteries connected in series, the greater the risk of overcharging and over-discharging. The performance of the battery is evaluated based on these parameters and characteristics [22]. Figure 10.2 shows the classification of lithium-ion battery parameters.

a. Normal parameters of the battery
   - Nominal voltage – This voltage acts as the naming voltage for a particular voltage source so that category of voltage source can be recognized, i.e., the 12 V nominal voltage of a battery means its output will be near 12 volts. It can be 11.5 or 12.5, but that does not mean it would be exactly 12 volts. This criterion is used for

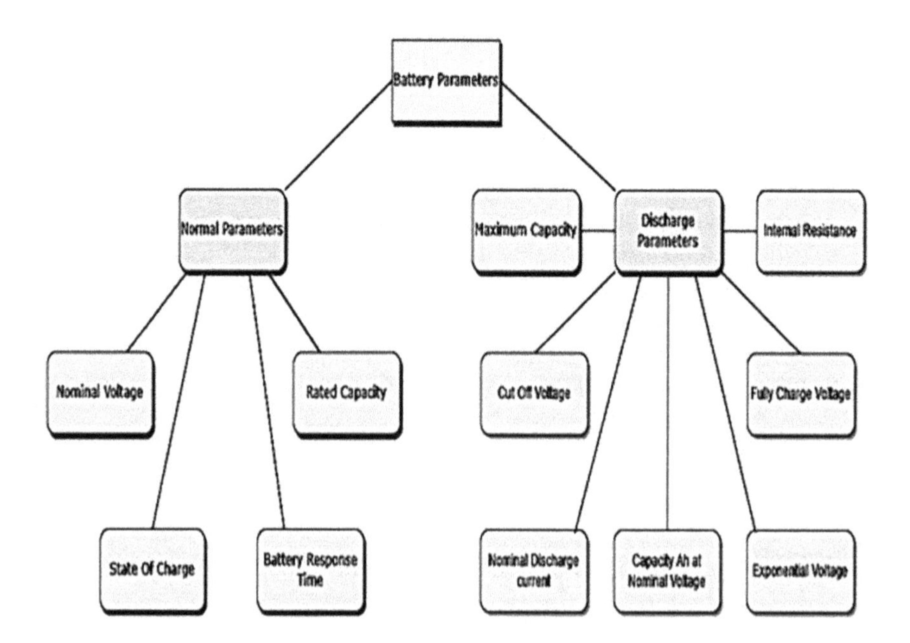

Figure 10.2 Classification of Li-ion battery parameters in MATLAB®.

battery marketing. Electrical power systems are named according to their voltage [23].

- Rated capacity refers to a cell's or battery's ability to supply steady current for a set amount of time. It is measured in ampere-hours (Ah). This capacity is determined by the battery/discharge cell's rate. The cell capacity decreases as the discharge rate increases, and vice versa.
- State of Charge (percent) – SOC refers to the battery's current capacity as a percentage of its maximum capacity. For determining the change in battery capacity over time, SOC is often calculated using current integration [24].
- Response time – The battery's response time is the value that represents voltage dynamic and can be observed when a current step is applied.

b. Discharge parameters of the battery

- Fully Charged Voltage – The voltage is achieved by the battery when it achieves full charge. It is the no-load voltage, i.e., a 12 V battery has 13.9678 V as a fully charged voltage. This voltage plays a major role in the determination of the overall service life of the battery [10].
- Maximum capacity (Ah) – It means maximal amp-hour capacity. Battery capacity is the total electricity generated from electrochemical reactions in the battery and is expressed in ampere-hours. Maximum capacity is defined as the theoretical maximum capacity when a discontinuity occurs in battery voltage and is represented by Ah [9].
- Cut-off voltage (V) – It is the minimum voltage allowed to be reached by a battery, below which the battery becomes inoperable. This voltage acts as a lower limit voltage at which battery discharge is complete and is the end of discharge characteristic. This voltage is usually chosen to achieve the max usable capacity of the battery [9].
- Nominal Discharge Current – The value against which the discharge curve is calculated. It is expressed as a percentage of C's numerical value. Discharge current can also be represented as a multiple of the rated discharge current [9].
- C-rate is the unit of measurement for a battery's charge and discharge current. The C-rate is a unit expressing a current value used to calculate and designate the battery's projected sufficient time under varied charge/discharge conditions. The majority of portable batteries have a 1C rating. If a 1000 mAh battery is discharged at a 1C pace, it will deliver 1000 mA for 1 hour. The identical battery would provide 500 mA for 2 hours if drained at 0.5C. The 1000 mAh battery would provide 2000 mA for 30 minutes at 2C. A 1-hour discharge is typically referred to as 1C; a 0.5C is a 2-hour discharge, and a 0.1C is a 10-hour discharge [25].

- A battery's capacity (Ah) at nominal voltage is defined as the number of amp-hours it can store. Qom represents it and is taken from the battery until the voltage drops below the nominal voltage. This number should fall somewhere between Qexp and Qmax. The total amp-hours available when a battery is depleted by a specific discharge current (given as C-rate) from 100% State of Charge (SOC) to Cut of Voltage (COV) is known as coulometric capacity [8].
- Exponential zone [voltage (V), capacity (Ah)] – Vexp, and Qexp, which corresponds to the end of the exponential zone, are known as exponential zone voltage and capacity. The voltage should be between Vnom and Vfull. Capacity should be in the range of 0 to Qnom [8].
- IR (Ohms) is the resistance to current flow inside the battery. The two basic components impacting a battery's IR are electronic and ionic resistance.

  A generic value of 1% of the nominal power is loaded when a present model is utilized. During charge and discharge cycles, the resistance remains constant and does not change with the current amplitude [11].

10.3 DISCHARGE PARAMETER ESTIMATION METHODOLOGY

Firstly, an appropriate battery was selected based on studies. All the parameters of the selected battery were identified. Then, the category of this battery was selected according to Figure 10.2. Based on the information available in the library, the data of a standard Li-ion battery was taken, and a simulation model of a Li-ion battery was developed [26]. Simultaneously, based on the data available on the internet, some practical batteries were also incorporated into this study. The discharge parameters have been selected as performance parameters. IR is not included in the work because its percentage deviation could not be found, and the available data is not sufficient. In this work, three Li-ion batteries with different ratings, 3.7V, 2.6 Ah; 6V, 10 Ah; and 12V, 20 Ah, are considered for this work, and the parameters of the same are presented in Table 10.2.

The whole process and calculation are shown in Figure 10.3 to calculate the percentage deviation using the switching algorithm (switch case). For this, Table 10.2 is taken as the input data, and the percentage deviation of each of the three cells is determined. For each of the three batteries, the percentage deviation of their parameters concerning their average rating is calculated and displayed in Table 10.3. This table helps to conclude that for any Li-ion battery, the discharge parameter varies with a certain percentage

Table 10.2 Li-ion battery parameters of three cells

Parameters								
Normal Parameters				*Discharge Parameters*				
Nominal voltage (V)	Rated capacity (Ah)	State of charge (%)	Battery response time (s)	Cut-off voltage (V)	Fully charged voltage (V)	Nominal discharge current (A)	Capacity (Ah) at nominal voltage	Exponential zone [voltage, capacity]
3.7	2.6	96	30	2.775	4.3068	1.1304	2.3513	[3.9974 0.12774]
6	10	96	30	4.5	6.9839	4.3478	9.0435	[6.4823 0.4913]
12	20	96	30	9	13.9678	8.6957	18.087	[12.9646 0.982609]

value, despite their different ratings. As seen in the case of all three batteries, i.e., 3.7, 6, and 12 V, there is a deviation from the full charge voltage (–16.40%). The negative sign represents the discharging nature. Figure 10.4 stipulates the common parameters value of any discharge parameters. The assisted discharge parameters from 48.1 V and 20 Ah are given as inputs for nominal voltage and nominal current, respectively, for calculation and the output is shown in Tables 10.3 and 10.4.

10.4 RESULTS AND OBSERVATIONS

Table 10.3 represents the various results obtained through the algorithm of percentage deviation (shown in Figure 10.3). The percentage deviation of each parameter is the same for every cell, which can be considered and their values will be fixed for further calculations. The obtained results of percentage deviation are shown in Table 10.3.

From this table, certain observations can be concluded, which are mentioned below:

- Fully charged voltage (V) deviates by –16.40% concerning that cell's nominal voltage; this deviation is the same for all three batteries that are taken into consideration.
- Nominal discharge current (Ah) deviates by 56.2% concerning that cell's nominal current; this deviation is the same for all three batteries that are taken into consideration.
- Cut-off voltage (V) deviates by 25% concerning the cell's nominal voltage; this deviation is the same for all three batteries that are considered.

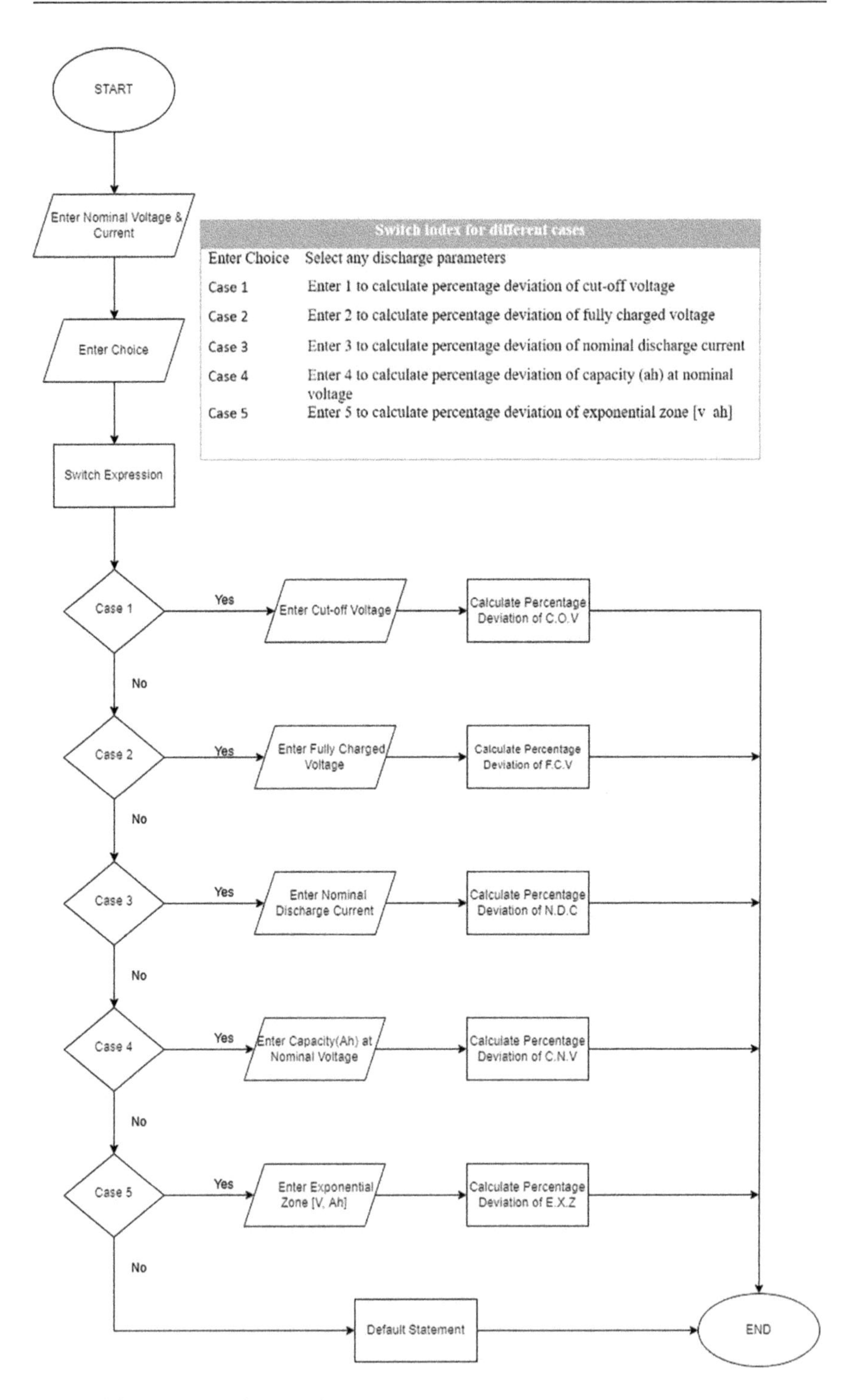

Figure 10.3 Flow chart for calculating percentage.

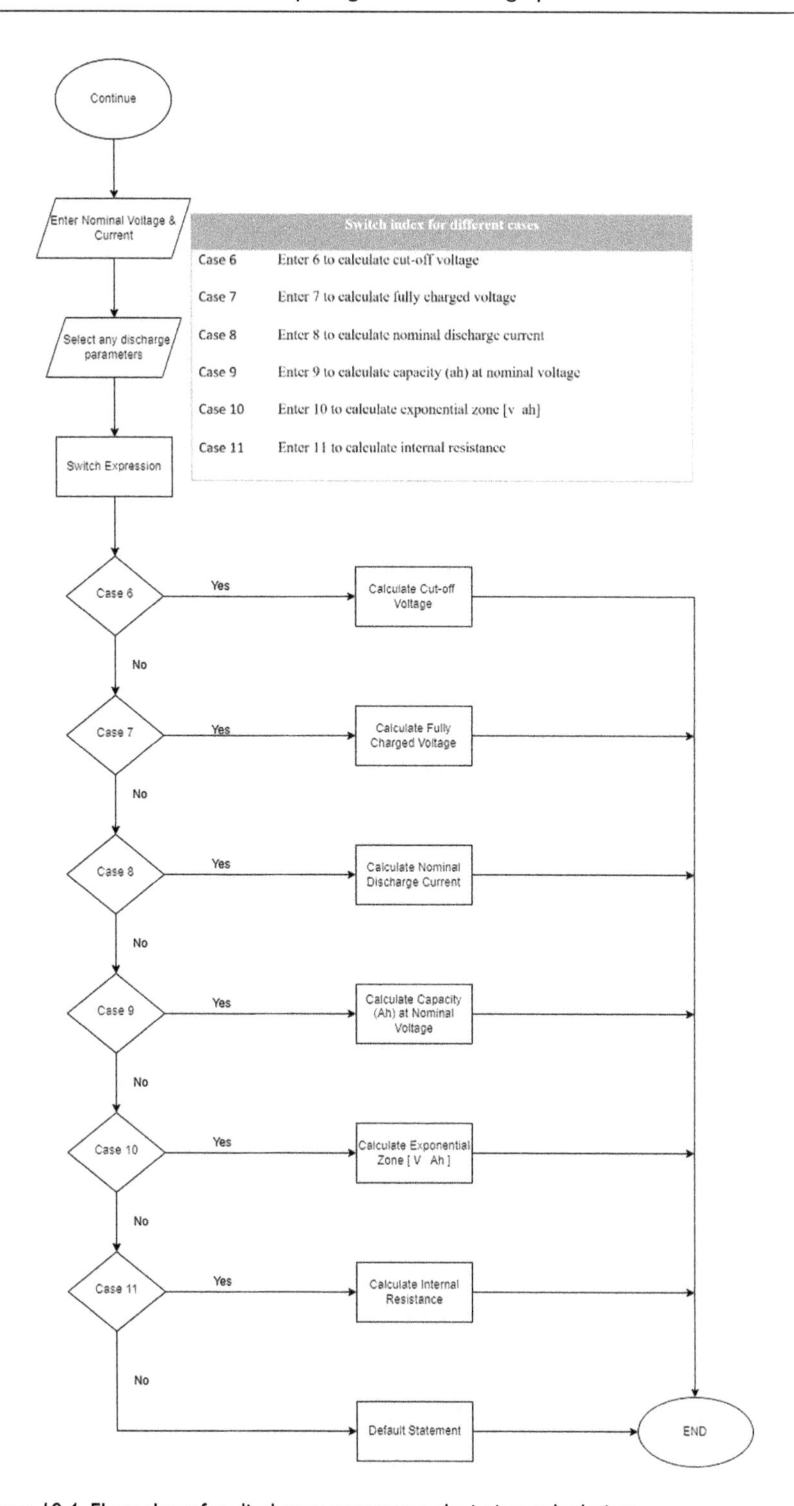

Figure 10.4 Flow chart for discharge parameter deviation calculation.

Table 10.3 Percentage deviation of discharge parameters

Parameters	*Batteries* 3.7V & 2.6Ah	6V & 10Ah	12V & 20Ah
Fully charged voltage (V)	−16.40%	−16.40%	−16.40%
Nominal discharge current (Ah)	56.52%	56.52%	56.52%
Cut-off voltage (V)	25%	25%	25%
Capacity (Ah) at nominal voltage	9.56%	9.56%	9.56%
Exponential zone [voltage (V), capacity (Ah)]	[−8.04%, 95.09%]	[−8.04%, 95.09%]	[−8.04%, 95.09%]

Table 10.4 Values of discharge parameters via normal parameters

Normal parameters			*Discharge parameters*				
Nominal voltage (V)	Rated capacity (Ah)	Cut-off voltage (V)	Fully charged voltage (V)	Nominal discharge current (A)	Capacity (Ah) at nominal voltage (Ah)	Exponential zone [voltage, capacity]	Internal resistance (Ohms)
11.1	10	8.325	12.9203	4.3478	9.0435	[11.9923, 0.491304]	0.0111
48	20	36	55.8714	8.6957	18.087	[51.8585, 0.982609]	0.024

- Capacity (Ah) at nominal voltage deviates by 9.56% concerning the nominal voltage of that cell, and this deviation is the same for all three batteries that are taken into consideration.
- Exponential Zone [Voltage(V), Capacity (Ah)] deviates by [−8.04% 95.09%] respectively, concerning nominal voltage and nominal current of that cell, and this deviation is the same for all the three batteries that are taken into consideration.

Table 10.4 represents the value of fully charged voltage, cut-off voltage, nominal discharge current, capacity at nominal voltage, exponential zone, and IR. Ohm's law can estimate the IR as Rint = V/I.

In this study, three practical batteries available in the market have also been included in applying the proposed estimation algorithm and validating it by obtaining results. These are Orange-18650 Li-ion Battery, Volta Li-ion Battery, and PPCB-4820 Li-ion Battery Packs, and the required parameters have been extracted from their specification datasheet. Analyzing the above batteries' data sheets shows that the proposed algorithm is estimated

Table 10.5 Li-ion battery parameters of three practical cells

Battery property	*ORANGE-18650 LI-ION BATTERY (11.1V & 10 Ah)*	*VOLTA LI-ION BATTERY (48V & 20Ah)*	*PPCB-4820 LI-ION BATTERY PACK (48V & 20Ah)*
Nominal voltage (V)	11.1	48.1	48.1
Rated capacity (Ah)	10	20	20
Series parallel configuration	3s 4p	3s 8p	N/a
Fully charged voltage (V)	12.6	54.6	54.5
Cut-off voltage (V)	8.25	37	35.75
Approximate weight (kg)	0.75	8	4.5
Battery dimensions (L×W×H) (in mm)	14×10×9	150×190×180	175×250×70

Table 10.6 Comparison of percentage deviation between practical battery and MATLAB® battery

	Battery			
	Practical battery available on the market			*Used in the present work*
Parameters (Unit)	Orange 18650 Li-ion battery (11.1V & 10 Ah)	Volta Li-ION battery (48V & 20 Ah)	PPCB-4820 Li-ion battery pack (48V & 20 Ah)	Battery analysis from Table 10.3
Fully charged voltage (V)	−13.51%	−13.51%	−13.31%	−16.40%
Cut-off voltage (V)	25.68%	23.08%	25.68%	25.00%

discharge parameters are almost correct. By taking values from Table 10.5 and using Figure 10.3, the percentage deviation of discharge parameter is calculated for practical batteries, and the result is shown in Table 10.6, along with the result also shows the comparison between the percentage deviation of practical batteries and the batteries used in the research. Table 10.6 shows that the average percentage deviation of fully charged voltage in the case of three practical batteries is –13.44%. By comparing this average (–13.44%) with the average percentage deviation of batteries used in the analysis (–16.40%), it is observed that the result provided by the algorithm shown in Figure 10.3 differs by 2.96%. At the same time, it has also been observed that the average percentage deviation of cut-off voltage is 24.81%. Comparing this average (24.81%) with the average percentage deviation of batteries used in the project (25.00%) differs by 0.19%.

10.5 CONCLUSION

In the present work, a software-based solution is provided for determining the discharge parameter of any lithium-ion battery. Percentage deviation has been evaluated based on available data and calculations. The datasheet of the discharge parameter can be estimated by taking as input the nominal parameters such as nominal voltage and nominal current. The evaluation has been done keeping in mind three different ratings of batteries. Even though the nominal rating of the battery may be different, the proposed algorithm can obtain a new datasheet relevant to the battery rating. Because it has been observed that the percentage deviation of the discharge parameter remains the same for any Li-ion batteries. The proposed estimation algorithm is justified by comparing the results with three different lithium-ion batteries available on the market. This work can also be modified so that the nominal rating of the battery at the output can be obtained by applying any number of discharge parameters. As a result, this function can be used for battery selection in any application area. Furthermore, evaluating the discharge parameters of a lithium-ion battery will not require any MATLAB® simulation or any external tools and the tedious task of finding the datasheet of a battery with the specified rating is simple. This algorithm can also be used in the future development of software-based estimators.

REFERENCES

1. Rahimi, M. "Lithium-Ion Batteries: Latest Advances and Prospects". Batteries 2021, 7, 8. doi: 10.3390/batteries 7010008.
2. Bellis, Mary. Biography of Alessandro Volta, Inventor of the Battery. About.com. Retrieved 7 August 2008.
3. Battery History, Technology, Applications and Development. MPower Solutions Ltd. Retrieved 19 March 2007.
4. Borvon, Gérard 10 September 2012. "History of the electrical units". Association S-EAU-S.
5. Brudermüller, Martin; Sobotka, Benedikt; Dominic, Waughray (September 2019). Insight Report—A Vision for a Sustainable Battery Value Chain in 2030 : Unlocking the Full Potential to Power Sustainable Development and Climate Change Mitigation (PDF) (Report). World Economic Forum & Global Battery Alliance. pp. 11, 29. Retrieved 2 June 2021.
6. Rahmoun, A.; Biechl, H., "Parameter's identification of equivalent circuit diagrams for li-ion batteries". In Proceedings of the 11th International Symposium PÄRNU "Topical Problems in the Field of Electrical and Power Engineering" and "Doctoral School of Energy and Geotechnology", Pärnu, Estonia, 16–21 January 2012.
7. Saldaña, Gaizka, José I. San Martín, Inmaculada Zamora, Francisco J. Asensio, and Oier Oñederra. "Analysis of the Current Electric Battery Models for Electric Vehicle Simulation" 2019, Energies 12, no. 14: 2750.

8. Kurniawan, Ekki & Rahmat, Basuki & Mulyana, Tatang & Alhilman, Judi. (2016). Data analysis of Li-Ion and lead acid batteries discharge parameters with Simulink-MATLAB. 1–5. 10.1109/ICoICT.2016.7571959.
9. C. Tang, T. Wang, T. Jiang, Y. Tang and J. Sun, "Remaining Discharge Capacity Online Estimation for Lithium-Ion Batteries Under Variable Load Current Conditions, " IECON 2020 The 46th Annual Conference of the IEEE Industrial Electronics Society, 2020, pp. 1911–1916, doi: 10.1109/IECON43393.2020.9254332.
10. T. M. Tuan and W. Choi, "Design and Implementation of a Constant Current and Constant Voltage Wireless Charger Operating at 6.78 MHz, " 2019 10th International Conference on Power Electronics and ECCE Asia (ICPE 2019-ECCE Asia), 2019, pp. 1–6, doi: 10.23919/ICPE2019ECCEAsia42246.2019.8797329.
11. D. Qian et al., "Research on Calculation Method of Internal Resistance of Lithium Battery Based on Capacity Increment Curve," 2019 2nd World Conference on Mechanical Engineering and Intelligent Manufacturing (WCMEIM), 2019, pp. 343–346, doi: 10.1109/WCMEIM48965.2019.00074.
12. Kaiyuan Li, King Jet Tseng, Lemuel Moraleja, "Study of the influencing factors on the discharging performance of lithium-ion batteries and its index of state-of-energy", IECON 2016–42nd Annual Conference of the IEEE Industrial Electronics Society.
13. Hardik Keshan, Jesse Thornburg and Taha Selim Ustun, "Comparison of Lead-Acid and LithiumIon Batteries for Stationary Storage in Off-Grid Energy Systems", 4th IET Clean Energy and Technology conference (CEAT 2016).
14. Koehler, U.; Kruger, F.J.; Kuempers, J.; Maul, M.; Niggemann, E.; Schoenfelder, H.H. (1997). [IEEE IECEC-97 Thirty-Second Intersociety Energy Conversion Engineering Conference (Cat. No.97CH6203) - Honolulu, HI, USA (27 July-1 Aug. 1997)] IECEC-97 Proceedings of the Thirty-Second Intersociety Energy Conversion Engineering Conference (Cat. No.97CH6203) - High performance nickel-metal hydride and lithium-ion batteries.
15. Andrew F. Burke, "Batteries and Ultracapacitors for Electric, Hybrid, and Fuel Cell Vehicles", Journals and magazines, Proceedings of the IEEE, Volume: 95 issue 4.
16. David Linden, Thomas B Reddy, "Handbook of Batteries", Third Edition, McGraw- Hill, 2001.
17. O. Tremblay and L.A. Dessaint, "Experimental Validation of a Battery Dynamic Model for EV Applications," World Electric Vehicle Journal Vol. 3, ISSN 2032–6653, AVERE, pp. 0289–0298, 2009
18. R. Benger, H. Wenz, H.P. Beck, M. Jiang, D. Ohms, G. Schaedlich, "Electrochemical and thermal modeling of lithium-ion cells for use in HEV or EV application," World Electric Vehicle Journal Vol. 3, ISSN 2032–6653, AVERE, pp. 0342–0351, 2009.
19. Li Wencheng, Lu Shigang, Pang Jing, "Preparation and performance of high power lithium secondary battery", Power supply technology, 2009, pp. 280–283.
20. Huang, Jian & Xu, Peichen & Wang, Peiyong. (2020). Experimental measurement of anisotropic thermal conductivity of 18650 lithium battery. Journal of Physics: Conference Series. 1509. 012013. 10.1088/1742–6596/1509/1/012013.

21. Shuai Ma, Modi Jiang, Peng Tao, Chengyi Song, Jianbo Wu, Jun Wang, Tao Deng, Wen Shang, Temperature effect and thermal impact in lithium-ion batteries: A review, Progress in Natural Science: Materials International, Volume 28, Issue 6, 2018, doi: 10.1016/j.pnsc.2018.11.002.
22. V.M. Dileepan, J. Jayakumar, "Performance Analysis of Lithium Batteries", Proceedings of IEEE International Conference on Innovations in Electrical, Electronics, Instrumentation and Media Technology ICIEEIMT 17.
23. Chang, C.-k. "Factors Affecting Capacity Design of Lithium-Ion Stationary Batteries. Batteries", 2019, 5, 58. doi: 10.3390/batteries5030058.
24. Xiaogang Wu, Xuefeng Li, And Jiuyu Du, "State of Charge Estimation of Lithium-Ion Batteries Over Wide Temperature Range Using Unscented Kalman Filter", Special Section On Advanced Energy Storage Technologies and their Applications, 2019, doi: 10.1109/ACCESS.2018.2860050
25. V. Sangwan, A. Sharma, R. Kumar and A. K. Rathore, "Estimation of optimal li-ion battery parameters considering c-rate, SOC and temperature, " 2016 7th India International Conference on Power Electronics (IICPE), 2016, pp. 1–6, doi: 10.1109/IICPE.2016.8079484.
26. Kim, J.; Kowal, J. Development of a Matlab/Simulink Model for Monitoring Cell State-of-Health and State-of-Charge via Impedance of Lithium-Ion Battery Cells. Batteries 2022, 8, 8. https:// doi.org/10.3390/batteries8020008.

Index

www.ingramcontent.com/pod-product-compliance
Ingram Content Group UK Ltd.
Pitfield, Milton Keynes, MK11 3LW, UK
UKHW021121180425
457613UK00005B/177